The Olive Oil Companion

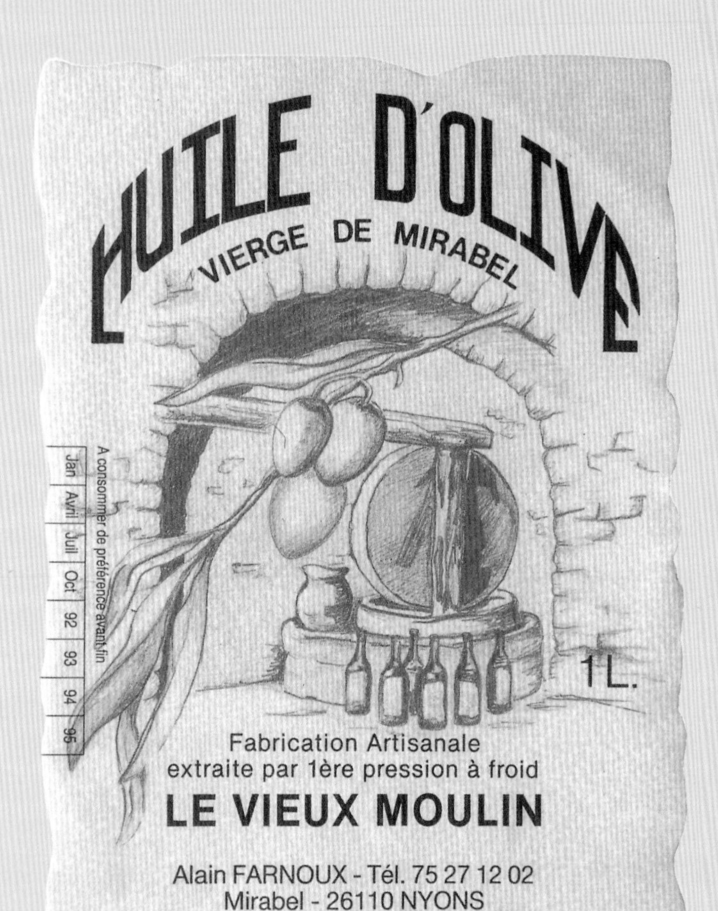
HUILE D'OLIVE
VIERGE DE MIRABEL
A consommer de préférence avant fin
Jan
Avril
Juil
Oct
92
93
94
95
1 L.
Fabrication Artisanale
extraite par 1ère pression à froid
LE VIEUX MOULIN
Alain FARNOUX - Tél. 75 27 12 02
Mirabel - 26110 NYONS

The Olive Oil Companion

The Authoritative Connoisseur's Guide

Judy Ridgway

A Quintet Book

Published by Knickerbocker Press
276 Fifth Avenue, New York, New York 10001

This edition produced for sale in the U.S.A.,
its territories and dependencies only.

ISBN 1-57715-005-8

This book was designed and produced by
Quintet Publishing Limited, 6 Blundell Street, London N7 9BH

Creative Director: Richard Dewing
Art Director: Patrick Carpenter
Designer: Roger Fawcett-Tang
Senior Editor: Anna Briffa
Editor: Anna Bennett
Photographer: Paul Forrester
Illustrator: Christine Sloane

Typeset in Great Britain by
Central Southern Typesetters, Eastbourne
Manufactured in Singapore by Bright Arts Pte Ltd
Printed in China by Leefung-Asco Printers Ltd

Contents

1

THE STORY OF OLIVE OIL
6

The History of the Olive
7

Olives Around the Globe
12

The World of Olive Oil
23

Appreciating Olive Oil
30

The World of Table Olives
36

2

DIRECTORY OF OLIVE OILS
41

Index
191

Author Acknowledgments

I would like to thank the many importers and producers who supplied me with copious amounts of oil to taste. Without their help it would not have been possible to write this book. I would also like to thank my editor, Anna Briffa, who coordinated the deliveries and generally smoothed the way.

THE STORY OF OLIVE OIL

1

The History of the Olive

The olive tree grew wild in the Middle East and its fruits have been used since prehistoric times. Its history is so ancient that no one knows who first pressed the olive for its oil or who thought of softening and preserving the fruit in salt or soda.

Along with the vine, the olive tree was one of the first plants to be cultivated and the practice spread from central Persia and Mesopotamia to Egypt and Phoenicia and then to Greece. By the dawn of history all the peoples of the eastern Mediterranean lived by the olive and the vine.

Above: Spanish olive groves; (left) the Siera de Segura, and (right) Jaén.

Because of its importance as a soothing anointment and as a source of both food and light, the olive gained a religious and divine significance. One of the earliest references to the olive occurs in an Egyptian papyrus from the twelfth century BC. It is a deed of gift from the Pharaoh Ramses III to the god Ra. In it he offers the olive groves which were planted round the town of Heliopolis: "From these trees the purest oil can be extracted to keep the lamps of your sanctuary burning."

Legend relates that a cedar, a cypress, and an olive tree grew on Adam's grave on the slopes of Mount Tabor. The Bible abounds in references to the olive. In Genesis it was an olive branch that the dove brought back to the Ark, proving to Noah that the waters were subsiding after the flood. Ever since, the olive branch has been regarded as a symbol of peace and goodwill.

Greek mythology describes how Zeus promised to give Attica to the god or goddess who offered the most useful invention. It was Athena, goddess of wisdom and peace, who won with her gift

of the olive tree and its soothing, nourishing oil. Athena became the goddess of Athens and her olive tree was said to be planted on the rock of the Acropolis. The story is commemorated in a frieze on the side of the Parthenon.

In the course of their travels the Greeks introduced the olive to Italy, where it soon established itself. The peoples of north Africa also cultivated the olive and it gradually moved along the coast through Tunisia to Algeria and Morocco and then north into Spain and Portugal.

In *De Re Rustica*, the Roman writer Cato tells his readers that every farm should place the olive high on its list of crops. The Romans brought their very practical skills to bear on the olive grove and its products and invented the screw press to perfect the method of extraction. They went on to improve the storage and distribution of oil.

Above: These huge terracotta jars, or Orci *as they are known in Italy, were traditionally used to store the olive oil after pressing.*

The Romans were also responsible for the further spread of olive groves, taking the trees into northern Italy until there was hardly a province which did not produce olives or olive oil. They also introduced the tree into Provence, but they still could not produce enough to fulfill their needs. Clay vessels of Spanish origin unearthed in Italy, stamped with the seals of the exporters, show that Spanish growers filled the gap, as they often do to this day.

In the oldest cookbook, *Reconquindaria*, which dates back more than 2,000 years, the author Apicius frequently refers to olive oil from Spain.

Though the trade in olive oil stopped with the fall of the Roman empire, the emphasis on the importance of the olive crop, along with that of the vine, survived to reassert itself in the monastic dominance of the Middle Ages. Oil was required for the illumination of the church and wine for the religious service.

In thirteenth-century Italy monks in the Salentino area of Apulia commenced widespread olive cultivation. They laid the foundations for the huge oil production of this part of Italy today. The Florentine region of Tuscany was also an important olive-producing region and Florence became the center of a great market for olive oil, establishing a reputation which it has really never lost despite its small production levels in modern times.

Venice, vying for trade with Genoa, developed special flat-bottomed ships equipped for the transport of jars. These boats transported oil produced in southern Italy back to the more heavily populated northern provinces. What was probably the first regulatory board for olive oil was established in Venice in the fourteenth century. The "Visdomini di Tenaria" were appointed to control olive oil imports and exports and to regulate weights and measures and retail trade.

The olive trade was so important to the southern Italian economy that after the Spanish conquest of the region in the mid sixteenth century, the victors ordered the construction of a road linking Apulia to Naples to ensure a more rapid transport of the oil for trade.

For centuries the production of olives and olive oil remained essentially a family business. In some areas cooperatives were formed to press or cure the olives but the production base continued to be quite small. All this changed at the end of the nineteenth century. The development of the industrial oil-refining plant meant that olive oil became just another commodity.

Some small centers of production remained in each country, producing the best extra-virgin oil or simply supplying a local demand, but vast quantities of olives went to the refinery to be made into anonymous olive oil which was then shipped around the world.

This trend might have continued until single farm or estate oils became a curiosity or a thing of the past. In the late 1970s,

Above: Olive oil is used throughout the Mediterranean for dressing salads and cold foods.

however, scientists in America and elsewhere began to realize the nutritional advantages of the so-called Mediterranean Diet in general and of olive oil in particular. Subsequent research appears to confirm the value of olive oil in the diet.

This regard for the healthy aspects of olive oil coincided with a revival of interest in quality foods produced on a small scale. In addition, many chefs in the U.K. and in the U.S. have moved away from the traditions of classical French cuisine toward a more Mediterranean approach.

Olives and olive oil have always been of interest to gourmet travelers but were not so well known outside their own countries. This is now changing and extra-virgin olive oil is appreciated by lovers of good food all over northern Europe and in America.

This growing interest in olive oil has led to a revival of quality oils and of single-estate production, particularly in Italy. However, this interest is not confined to Europe. Though most of the world's olive oil is produced around the shores of the Mediterranean, olives have been grown in other areas of the world for many years.

The explorers, travelers, and migrants of the past took the olive with them and the tree flourished in those parts of the world which have a Mediterranean-type climate. Olives have been cultivated in California, for example, for more than 150 years. They were first introduced by Franciscan monks from Spain. Mission olives, named after the missions where they were first introduced, were joined in the 1870s and 1880s by a number of other European olive varieties, and press houses sprang up all over the state.

Canned olives, which began to be marketed at the turn of the century, killed the Californian olive oil industry – canned olives were more profitable than olive oil, and cheap European oils flooded the market. Many olive groves were abandoned and became overgrown with tangles of manzanita, oaks, and brush.

In other areas such as Lindsay in the Central Valley olives had merely been grown as windbreaks for the citrus groves but they were known to do well there. With the advent of canning, the introduction of a new black-ripe variety of olive, and a knowledge of curing techniques the area eventually became the center of the Californian table olive industry.

In recent years the growing numbers of educated and demanding consumers in the U.S. have brought about a resurgence in the production of Californian olive oil. Some wineries now grow olives as well as grapes and bottle "single-estate" oils and the number is increasing every day.

Olive groves were established by the missionary fathers in Mexico and the Argentine and, rather surprisingly, they also reached Australia more than a hundred years ago. In more recent years they have turned up in South Africa and in New Zealand.

Each of these countries now has a band of pioneering olive growers who are experimenting with different clones to suit different soil types and trying out modern methods of cultivation and production.

Olives Around the Globe

Cultivated olives belong to the botanical group *Sativa* which, with its wild cousin *oleaster*, is a sub-species of *Olea europaea*. They flourish in a Mediterranean-type climate with hot dry summers and cool winters. They are sensitive to cold during the growing period but need temperatures close to zero in the winter to induce a dormant state in which they can rest.

Olives can withstand freezing temperatures for short periods providing the thaw is gradual, but if the temperature falls below 50°F, as it did in Italy in 1984, the trees will die. They are less sensitive to heat and drought and can withstand quite long periods without rain but they will not produce as many olives. A long drought in Spain led to a greatly reduced harvest in 1995–96.

In the northern hemisphere the Mediterranean climate occupies a band between the 30th and 45th latitudes and this is where most of the world's olive trees grow. However, they also grow well at similar latitudes in the southern hemisphere. These latitudes correspond very much to those which are suitable for growing vines and olive groves and vineyards will often be found side by side.

Above: Olive trees are planted within the vineyards of the Pasolini Dall'Onda estate in Tuscany, Italy.

Above: Olive trees have the ability to grow on some of the roughest terrains.

Like vines, olive trees are not particularly demanding of the soil in which they grow and this means that olives can be grown on poor stony ground which is unsuitable for other crops. In certain parts of Spain, for example, olive trees are all that you will see once the mountains become too high for orange groves or vineyards. Olives can grow at altitudes of up to 1,400 feet. However, they will also thrive on good, well-drained soils at low altitudes.

The Olive and Its Fruit

If you have never seen an olive tree look for an evergreen tree with leaves that are dark green on the top and covered with much lighter silver scales on the underside. It was this silvery coloring, shimmering in the breeze, that attracted the attention of the artist Van Gogh. Fascinated by their color, he painted 19 pictures of olive trees.

The trees are slow to grow, taking four or five years to yield their first fruits and another ten to 15 to reach their full capacity. Once established, however, the olive tree can live for many years. There are stories of trees which have stood for 1,000 years. Some trees are known to have been around for 100 years or more, but really old trees are more likely to be the result of new shoots springing up from root systems which have survived the ravages of age or bad weather.

The unripe olive fruit is pear-shaped and green in color, changing to dark purple or black as it ripens. All olives, if left on the tree, will follow this pattern. Green table olives are picked and cured before they have ripened. Others are left on the trees and picked when they are fully ripe.

Olives which are to be pressed for oil may be picked at any stage in their development but the yields from unripe olives will be very small and the oil can be very bitter. Most olives are therefore left to ripen fully.

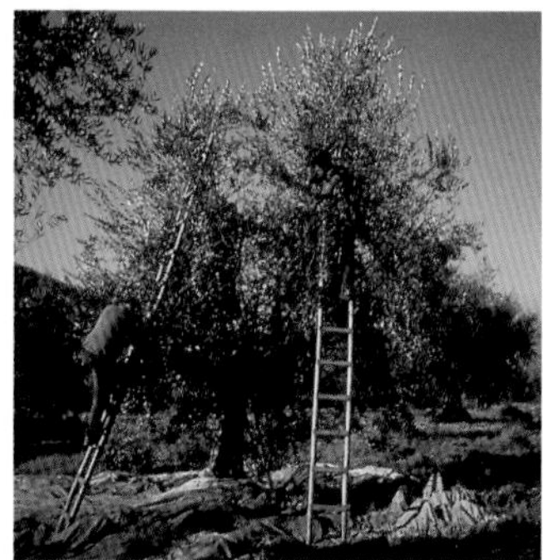

Above: Olive picking in progress on the Frescobaldi estate in Tuscany, Italy.

Cultivation and Harvesting

There are at least 50 or more different varieties of olive, each with its own distinct characteristics. Some olives, such as the Spanish Picual, are particularly suited to the production of oil, others like the French Lucque make better table olives. Some, like the Italian Frantoio, produce a hot peppery oil, while the Italian Taggiasca gives a much softer and sweeter oil.

Most olive-growing areas have their own particular varieties of olive, some of which do not grow outside that area. Very often the oils are produced from a random mix of these varieties. However, some of the more sophisticated growers deal with their olives in much the same way as they treat their vines. These people grow the different varieties in separate olive groves, then press and bottle them separately. Alternatively they may blend the different oils to produce a consistent flavor each year.

Everything that happens to the olive tree from pruning in spring through flowering, fruiting, and harvesting in the late fall will have a bearing on the quality of the fruit and thus on the

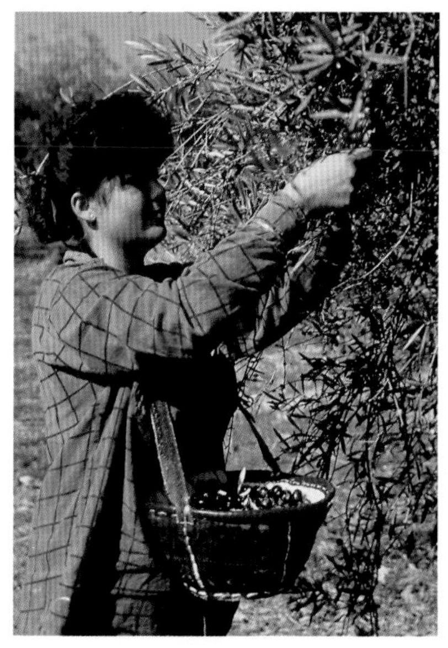

Above: Harvesting olives for the Nunez de Prado oil in Andalusia, Spain.

product it will yield. In many areas the methods used are traditional to that region.

The harvest is an extremely critical time as far as ripeness is concerned. Most growers want to produce as much good-quality oil as possible and this means optimum ripeness, but if the olives are left on the trees for too long they will overripen and oxidize as soon as they are picked producing unpleasant oil.

Freshly picked olives seem quite robust and hard but in fact they are very easily damaged so they must be handled with great care. In the wealthier groves they are harvested by hand. The pickers climb ladders and comb the olives from the trees with wooden rakes. Nets are stretched just above the ground to catch the olives as they fall.

In poorer areas the olives are left to fall as they ripen and the nets are emptied from time to time. This can cause problems because the time spent between picking and pressing is very important. In hot climates olives will start to oxidize as soon as they are picked.

In cooler climates the problem is less acute. Growers may deliberately leave their olives to stand for a day or two outside the mill. The olives start to heat up a little and so produce more oil. If this is carefully done it does not affect the quality of the oil.

Single-Estate Oils

A good deal of the olive crop in each producing country goes to large cooperatives or industrial plants to be processed into olive oil of various grades and into table olives. Some of these establishments have built up a name for good-quality produce.

The rest of the olives are pressed or processed on farms or at small local cooperatives. The quantities may be such that the oil is simply used by the grower, his family and friends, and perhaps his neighbours.

Other farms and estates produce somewhat larger quantities of oil which they press and bottle on the premises and sell as 'single-estate' oils. In recent years these producers have been able to establish a reputation for very high quality which allows them to charge top prices for their oils. As more people come to appreciate the virtues of first-class olive oil single-estate oils are finding a growing market and are increasing in number.

Because there is such a close affinity between vines and olive trees and between wine and oil, a good many of the single-estate oils come from farms and estates which also produce wine. This is particularly noticeable in central Italy and increasingly in California.

Spain

Spain is a hot, dry and dusty country with vast open spaces which are, in most instances, a long way from the sea. The heat builds up during the summer months and beats down on the olive groves resulting in a huge olive harvest. Indeed Spain often tops world production figures.

Olive trees can be found in every region of Spain and in most places olive oil and table olives are equally important. Outside Spain the best-known area for olive oil is probably Lérida, yet this region only accounts for about one per cent of Spain's total olive oil. On the other hand, Jaén, which produces more than a third, and Córdoba, which accounts for a further eighth, are much more important.

Most olives are grown by small farmers and taken to factories like Carbonell and Ybarra or to one of the many cooperatives for milling and pressing. However, in some areas such as Baena and Sierra de Segura there are some single estates which concentrate on producing very high-quality oil.

A very large number of olive varieties are grown in Spain and it is difficult to separate the different varieties from those which go by alternative local names. Among the most important oil-producing olives are the Arbequina and Verdial of Cataluña and Tarragona and the Picudo and Picual of Córdoba, Granada and Jaén. Other widespread varieties are Hojiblanca and Lechin.

The olives are mainly hand-picked and, as elsewhere, hydraulic presses, and more recently, centrifugal processes have almost completely replaced the old screw and lever presses. In some areas two types of oil are produced, one from young, early-picked olives and the other from mature olives picked later in the season.

Over the last few years efforts have been made to cut the time between harvesting and pressing and this policy is resulting in Spanish oil of a much higher quality.

Table olives are particularly important in the south. Here the Manzanillo and Gordal olives are processed for the table.

ITALY

Italy vies with Spain as the top olive-producing country. Its hilly, often mountainous countryside is ideal for the cultivation of olives and there is only one province in the entire country which does not produce them. This is the Val d'Aosta in the northwestern corner of Italy.

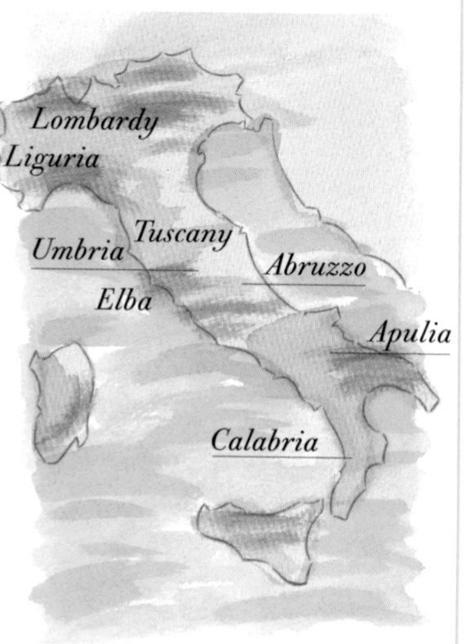

However, it is Tuscany which has built up the highest reputation for first-class olive oil. This is partly due to the fact that it produces more extra-virgin oil than any other region and partly to clever marketing. The region actually only accounts for a meager two to three percent of Italy's total production.

Outside of Italy, Lucca is the best-known name in the area, but very good oils are also produced in the Florence area and Colli Senesi. Many of the Tuscan groves were badly damaged in the frosts of 1955 and 1984 and they are just beginning to recover.

Oils from Liguria and Lake Garda in the north are also beginning to gain in popularity. But it is southern Italy which is the great producing region of Italy. Apulia and Calabria between them produce around 70 percent of Italian olive oil. This is backed by 10–11 percent from Sicily and five from Campania. In the past oils from all these areas were blended with other oils and sold under brand names but there is now a move by growers to produce a better-quality oil, marketed under their own names.

Despite its high levels of production Italy is a net importer of olive oil. This means that not all the oil exported from Italy is of Italian origin, even though the label is Italian. The new European Union Denomination of Origin laws should help here but they still only deal with top-class oils.

Like Spain, Italy has a very large number of olive varieties. Broadly speaking, the Coratina is extremely important in the south, Frantoio, Leccino, and Moraiolo dominate the central regions, and Taggiasca rules in Liguria.

Greece

It will come as no surprise to those who have vacationed in Greece that the Greeks consume more olive oil per capita then any other people in the world. They are also the third largest producer.

The mountainous region in the west of the country has been the traditional home of the Greek olive. Here small farms produce crops of 800–1000 pounds. The olives are taken to the cooperatives where they are mixed together and sold as blends. Today more groves are being planted in the plains of central Greece.

The Greek islands are also extremely important olive oil producers. Crete has been producing good-quality oils since the Minoan age and Lesbos also supplies oil to the mainland.

There is a tendency to think that Greek olive oil is somehow inferior to other oils but the truth is that the best oils have traditionally been sold to Italy where they are very highly regarded. Until very recently the Greeks have not tried to market their oil as a premium product and so presentation and packaging have been minimal. There are very few single-estate oils.

The most famous Greek olive is the Kalamata. It has a particularly good flavor and is usually processed for table use. Very occasionally it is also pressed to make oil.

One of the oil-producing varieties is the Koroneiki and confusion often arises because one of the main growing areas for this variety on the Peloponnese is also called Kalamata. Thus "Kalamata" olive oil usually means Koroneiki oil from Kalamata. Mani is a neighboring oil-producing area and the two compete with each other to produce the highest-quality oil.

France

France is one of the smallest olive-producing countries but because of its excellent climate both its table olives and olive oil are of a consistently high standard. The olive groves are concentrated in a thick band along the Mediterranean coast from Spain in the west to the Italian Riviera in the east. They also spread up the Rhône Valley as far as Valence.

One of the main areas of olive-oil production is Nyons in northern Provence. Here olive groves vie with vineyards and lavender fields to form an attractive mosaic on the hillsides. Another important area is the dramatic Vallee des Baux, but high-quality oils are also produced in the Vaucluse, Gard, and Alpes Maritimes. If you want to catch the harvest visit the region in early December.

Table olives are much more important in the Languedoc/Roussillon area with centers at Nîmes and Montélimar. They are also produced in Nyons and in Nice.

Unlike some other areas, France only has a small number of olive varieties, most of which are used to produce both olive oil and table olives. They include La Tanche – the predominant variety in Nyons – Picholine, Grossane, and Salonenque. Lucque (Languedoc) and Cailletier (Nice) varieties are more important for table olives than oil.

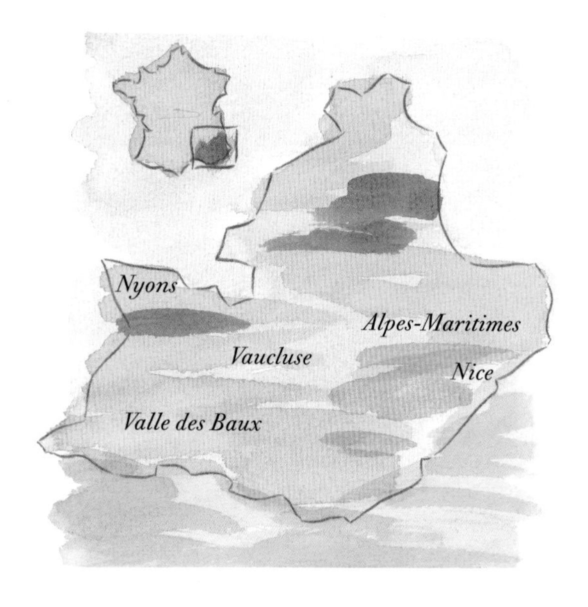

America

California is the center of the American olive industry, though olives are also grown in other parts of the United States such as Oregon, Washington State, and even Texas. Until recently the main thrust of the business was in producing table olives. Cheap olive oil was also made by gathering unpicked table olives or olives which were deemed unsuitable for pickling.

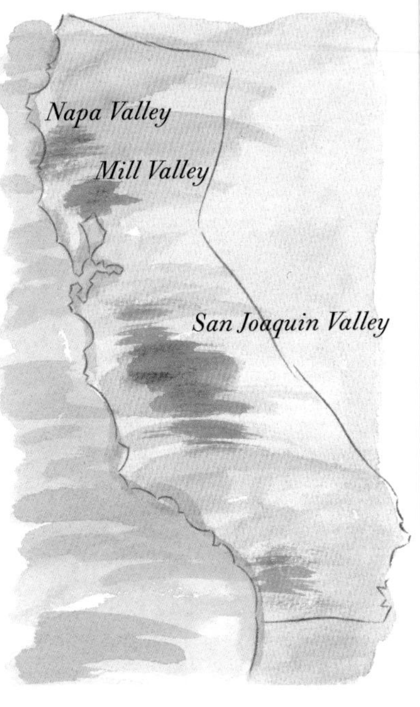

The most common olive is the Mission olive, brought in by the Franciscan fathers, followed by Manzanillo and Sevillano. All can make reasonable olive oils but the latter two are large in size and have a low oil yield. They are really much better suited to the production of table olives.

Since the late 1980s there has been a quiet revolution in the approach to olive oil production in California. A pioneer group of farmers have planted new olive groves and are endeavoring to produce top-quality oils to rival those of Europe. Many of these new producers are associated with wineries.

Most growers have chosen to introduce Italian olive varieties including Frantoio, Leccino, Pendolino, Moraiolo, and Taggiasca. These olive groves are still quite young and have not yet reached their full potential. Production is small but it is growing.

The success of these pioneers has attracted more growers into the field and California looks set to expand its production of premium oil.

The Rest of the World

Tunisia is the next largest producer of olive oil in the world after Spain, Italy, and Greece and by far by largest producer in north Africa. Oil is very important to the Tunisian economy as more than 20 percent of the population is engaged in its production. Half of the oil is exported to non-European Union countries.

Morocco and Algeria also produce olive oil and along with the countries of the Middle East, Libya and Argentina account for about seven percent of world production. Turkey, too, is an important player in world markets.

Portugal has as many trees as Tunisia but the deeply mountainous terrain is so barren that the average production of oil per tree is not as high as that in other countries. Nor is the quality particularly startling. This may be due to the fact the groves are not considered to be of particular importance and little attention is paid to careful production techniques. The weather is usually very hot at the harvest and the olives often overheat before they are pressed.

Most of Portugal's olive harvest is consumed at home and the Portuguese seem to be happy with their particular style of oil. The best oil is said to come from the upper Douro where port is produced. Some of the great port houses also grow olives but they rarely bottle and market their oil for export as the leading Tuscan estates do in Italy.

Production outside the countries listed above is so small as to be virtually negligible. However, this may change as olive oil and olives grow in popularity. Australia, New Zealand, and South Africa might well follow California's lead in small-scale production of high-quality oils.

Above: A stunning view across the Sierra Magina, Spain.

The World of Olive Oil

Olive oil is unique among vegetable oils in that it is produced, by purely mechanical means, from the fresh flesh of the fruit. In many places the simple process of grinding or milling the olives and then pressing them has changed very little since Roman times – the equipment has simply become more sophisticated.

Processing Olives for Olive Oil

Some mills still use huge granite millstones, others use modern stainless steel units and, in most parts of Europe, the hydraulic press has replaced the old screw press. This has had an enormous effect on the yield of oil and on its quality.

Above: The original mill at the Tenuta di Forci estate in Lucca, Italy.

The old presses only managed to extract about 40 percent of the oil. This was known as the "first pressing" and the oil was often labeled as such. Sometimes it was also labeled "cold-pressed."

The next step was to add hot water to the paste and to press it again to produce the "second pressing" or "hot-pressed" oil – terms which rarely appeared on the label! The modern hydraulic

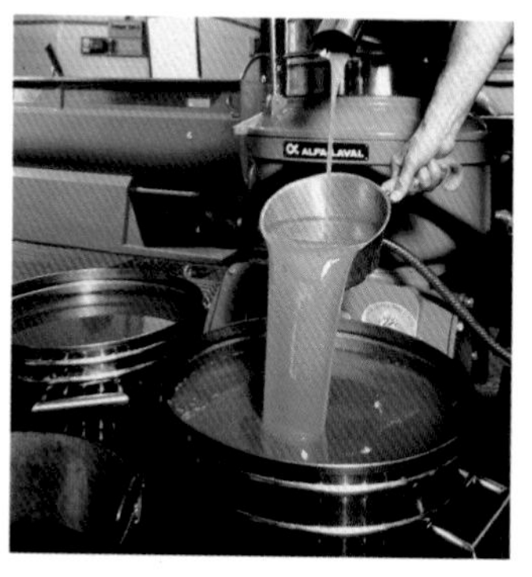

Above: The modern centrifugal plant at the Frantoio Gaziello company, Liguria, Italy.

press extracts more than 90 percent of the oil and the remaining pomace is then sent to the refinery for further processing. Thus there is no second pressing or hot-pressed oil. Some producers still like to put the old phrases on the labels of their bottles but they have no meaning today.

The fruit is first separated from leaves and twigs and then washed. Two or three millstones crush the fruit and its pits to a paste. The pits are important: the broken parts help to channel the oil when the paste is pressed. The milling process continues for about half an hour. During this time the cells of the fruit start to break down and release the oil.

The paste is then spread evenly over small round woven mats which are piled up in batches of 30 or 40 on the hydraulic press. The paste cannot be pressed in one great mass because it is very elastic and would resist the pressure. The mats are designed to allow the oil to trickle out and down the stack and collect at the bottom of the press.

The presses produce a reddish-brown liquid which is part oil and part natural olive vegetable water. The two are separated in a centrifuge. The oil is now immediately stored in underground tanks which will keep it fresh for quite a long time. In the past this process was carried out by slowly decanting the oil into troughs. The oil was then skimmed off as it rose to the surface. Some farms still like to use this method today, in which case the oil may be labeled Affiorato.

The pomace left on the mats from the hydraulic presses and the olive water will usually be sent to a refinery, where the last ounce of oil will be extracted for industrial use. Sometimes it is fed to cattle or put on the land.

A new phrase which some producers like to include on their label or in their promotional literature to indicate quality is "traditional method." This refers to the above method of production as opposed to a new process of separating the oil directly from the paste by centrifugal methods. This may be used after the initial traditional grinding with granite millstones, or as part of a continuous factory process which is still purely mechanical. Provided that a mill is hygienically operated there is no difference in quality between oils produced by one method or another. Nor is it possible to taste the difference.

However, if bad practices are allowed to creep in either method can result in poor-quality oil. Olive oil which does not reach the very highest standards is sent to the refinery to be refined and cleaned.

A few small single-estate producers use only the first-run oil which is generated purely by the weight of the olives themselves. The mills are organized in such a way that it is possible to extract a small amount of free-run oil before the paste is pressed. In Spain this oil, which is of an extremely high quality, is called *Yema flor* or yolk flower. It is bottled by hand, sealed with wax, and given a numbered label. The price, of course, reflects the extra effort!

In other areas the olives are pressed in a special press which does not use pressure. Instead the press works on the principle of percolation, exploiting the difference in the surface tension between oil and water. The olive paste is passed over thousands of shifting stainless-steel blades. The oil adheres to the blades and is funneled off. Here, too, the result is an oil of exceptional purity.

Grading Olive Oil

Around ten percent of the olive oil produced in the world is top-quality virgin olive oil. The rest has to be refined to remove impurities which affect the flavor and aroma of the final oil.

Originally the only way to judge olive oil was by tasting it. This was often done by a single person and was very subjective. Tasting remains important but today chemical tests have been developed to test the acidity levels of olive oil. The lower the acidity the better the oil.

Acidity in olive oil cannot be tasted as it can in lemons or vinegar. You can only taste this kind of acidity if there is something very wrong with the oil and it has gone rancid.

Extra-Virgin Olive Oil

This is the top grade of olive oil. It is a virgin, or unprocessed, oil which must have an acidity level of not more than 1g per 100 g or one percent. Some extra-virgin oils have acidity levels even lower than this. The oil must also have perfect aroma, flavor, and color.

Virgin Olive Oil

This is a virgin oil which must have an acidity level of not more than two percent, still with a perfect aroma, flavor, and color.

Olive Oil

This is a blend of refined and virgin oil. It must have an acidity level of not more than one and a half percent. Refined oil has no taste or smell and virgin oil is added to give it flavor. The oils vary in the amount of virgin oil that is added to them and so in the concentration of their flavors.

Olive Pomace Oil

This is refined oil which is extracted from the olive pomace that is left in the hydraulic press or centrifuge after the majority of the oil has been squeezed out. The oil should have an acidity of not more than one and a half percent. Like ordinary olive oil it is flavored with extra-virgin oil. Under no circumstances should it be labeled "olive oil."

Labels of Origin

Certain small regions produce olive oils of such exceptional quality that they have been given their own labels of origin. Spain was the first country to establish such a d.o.c. system and they still lead the way in this respect.

The Instituto Nacional de Denominaciónes de Origen in Spain recognizes a number of fully established d.o.c. areas with more pending. The established regions are Baena and Sierra de Segura in Andalusia and Les Garrigues and Siurana in Cataluña.

France followed Spain's lead, awarding an Appellation d'Origine Contrôlée or a.c. to Nyons olive oil and Nyons table

Above: Label of origin from the Baena region of Spain.

olives. The European Union introduced a union-wide D.O.C. system in 1995–96 and more D.O.C. areas are expected to enter this, particularly in Italy.

The D.O.C. or A.C. system for olive oil is very similar to that for wine. The oils must come only from within the designated areas and must be made from approved varieties of olive in the correct manner.

Styles of Olive Oil

The definitions for virgin olive oils say that the oil should have a perfect aroma, flavor, and color, but this does not mean that they taste the same. On the contrary, there is an enormous range of tastes and flavors.

Styles vary from the very sweet and mild to the very bitter and pungent. The oils also vary in how peppery they are. It is sometimes assumed that good olive oil should be very peppery but this is not the case. There are some very excellent oils in all styles and the choice is a matter of personal preference.

To some extent the overall style of an oil will depend on its place of origin. However, it would be wrong to assume that all the oils from one area taste the same. They do not. Tuscan olive oil, for example, is generally perceived to be green and pungent with a good deal of pepper, since these are the attributes you would expect from an oil which is produced from early-picked olives with Frantoio in the mix. Many Tuscan oils do indeed taste just like this, but there are others which are smoothly luscious, nutty, or even sweet. It would be quite easy to line up half a dozen or more oils which do not conform to the generally accepted style. Conversely, it would be equally easy to find oils from other parts of Italy or even outside Italy which taste as you might expect a Tuscan oil to taste. This is true of all olive growing regions, where there will always be exceptions to the norm.

Not only that, but even among those oils which do adhere to the expected flavor pattern there is still a good deal of variation. All this adds to the fun and enjoyment of discovering olive oil. It also means that you should read the following style guide with a good deal of caution, reserving judgment on any oil you buy until after you have tasted it! Indeed, if it is possible, it is better to taste before you buy.

Spain

Overall, Spanish oils tend to be rich and nutty but they can be extremely fruity or even grassy. They vary in how peppery they are.
Lérida and Borjas Blancas: The characteristic flavor of these oils is nuts – often toasted nuts – or even nutty chocolate. They are sometimes slightly bitter with a peppery but sweet finish to them.
Siurana: Though situated quite near Lérida and using much the same olives, the flavors of these oils are quite different. Tomato skins and grass are the key features and they tend not to be very peppery.
Baena: These oils are characteristically extremely fruity. As well as a lovely fresh olive taste they are full of the flavors of melons and tropical fruits like passion fruit and bananas. They are smooth and sweet. Some are fairly peppery, others are not.
Jaén and Sierra de Segura: The flavor of these oils is fruity and aromatic with a slight bitterness.

Italy

Italian oils exhibit every style from sweet fruity oils with little or no pepper, through nutty oils with a punchy finish to very pungent oils with fiery pepper.

Liguria and the North

Ligurian oils tend to be mild and lemony-sweet with almond overtones. They can be quite peppery but tend toward the medium rather than the strong. Some oils are fresher, with apple and tomato aromas. The oils of Lake Garda and the Trentino vary in style from light nutty oils similar to those of Liguria to much more grassy, pungent flavors.

Tuscany and Central Italy

You will find almost every style of oil here. There are richly fruity oils with little pepper, nutty oils with chocolate overtones, and medium pepper and the kind of green and grassy oils which are thought to be typical of Tuscany alone.

Some producers carry the pungent green approach to great lengths, picking their olives before they begin to ripen very much at all. The result is an almost electric green color and a flavor of bitter salad leaves such as rocket. These oils are often extremely peppery as well.

This central area takes in Umbria, Abruzzi, Molise, and Sardinia. The oils from Sardinia have recently begun to win prizes at the top level. They are full and lush, but elegant, with plenty of sweet fruit and light to medium pepper.

Apulia and the South

Here, too, you will find all kinds of oils but the predominant style is thick and nutty with a strong and peppery kick. Some oils are much more herbaceous and fruity – a style which rather surprisingly is typical of some of the best Sicilian oils.

Greece

Greek oils are characteristically very grassy. They often smell of newly cut lawns. They are often rather plain but still very attractive. The degree of pepperiness varies from oil to oil.

France

French oils are very mild and sweet with fruity flavors and aromas varying from apples and pears to lemons and tomatoes depending on the origin of the oil. They are not usually very peppery but there are exceptions. Les Baux, for example, can produce quite grassy and very peppery oils, which are not at all what one would expect from this area since the oils are usually lemony smooth.

California

There is some variation in the style of Californian oils though they tend toward the mild and sweet rather than the pungently peppery. However, this could change as more of the Italian olive varieties are introduced.

Appreciating Olive Oil

All you really need to taste, appreciate, and assess olive oil is your sense of smell and taste. After all, if you enjoy eating good food you unconsciously assess flavors as you eat and make choices for the future.

If you like olive oil you will know that different oils can have very different flavors. Unlike wine, very little information is ever given on the label and so you need to rely on your own tastebuds as your only guide.

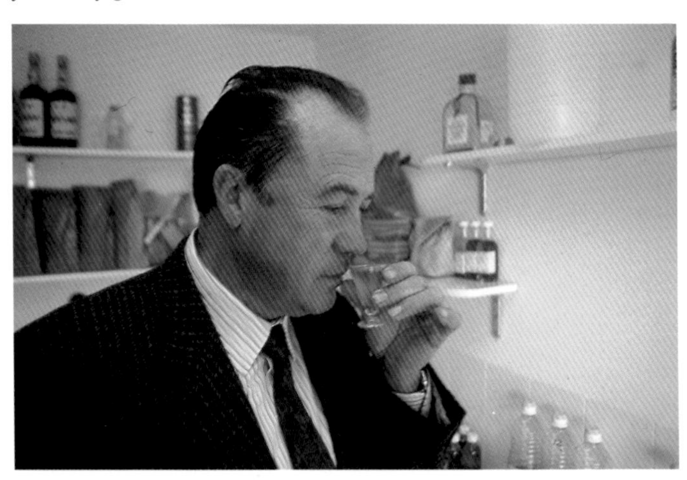

Above: Avelino Vea, of the Vea company, Lérida, Spain. He is one of very few recognized for his tasting abilities.

Some people think that the color of the oil will tell you something about its style. In fact, this is a very precarious way in which to judge an oil. Some oils with a strong green color do indeed have a "green" and punchy flavor but others do not and even if two oils both fall within this category they may still taste quite different.

Equally, pale yellow oils may be sweet and mild but they can also be very peppery. You really cannot tell from the color. In addition, good-quality oils are often packed in glass bottles of a dark color. This is to protect the oil from the light, to which it is sensitive.

Taste and Flavor in Olive Oil

Smell is the most important sense in tasting. If you have a bad cold and your nose is congested you will not be able to taste anything very much. Your sense of taste enables you to detect bitter flavors. It is the mucous membranes at the back of the throat which pick up the hot, peppery flavors. Your mouth is also very important in assessing the texture of an oil.

So what should you be looking for when tasting an oil? The first impression will be the overall style of the oil. Is it sweet and mild or strong and punchy? The second will be the degree of pepperiness. It is worth remembering here that the sensation of pepperiness increases as you add hot flavors to the palate. Thus the third or fourth oil you taste may seem more peppery than the first even if they are both at about the same level. If you taste the oils again in the opposite order you may decide that it is the first oil which is the most peppery!

Whatever olive oil snobs may say, there is no right and wrong in styles of oil. Nor is the oil of one region better than that of another. It is all simply a matter of personal preference. For example, punchy, peppery oils have no intrinsic value over lighter oils. However, there may well be a culinary difference. Very strong oils will kill the flavor of delicate foods and light oils will not be discernible in stronger dishes.

Once you have determined the style of an oil you will need to think about its taste and flavor. Olive oil is not just a cooking medium, it is also a flavoring ingredient in its own right. Thus you may want a particular type of flavor to match or contrast with the food being dressed or cooked in the oil.

Good olive oil should indeed taste of fresh olives but it may also have all kinds of secondary flavors which add to the enjoyment of it. The easiest way to describe these flavors is by reference to other foods. This does not mean that the oil tastes exactly like these other foods but that the aromas and flavors are reminiscent of them.

The range of flavors associated with olive oil is very wide. Fruit and vegetable tastes abound and include lemons, apples, tomatoes, pears, avocado, and almonds which are common. Verdant aromas like grass, new-mown hay, and salad leaves are also abundant. Tasters have also been known to come up with more unusual descriptions such as chocolate, charcuterie, passion fruit, and aniseed.

Cooking With Olive Oil

Olive oil is the traditional cooking medium of the Mediterranean. It is used in the way people from more northerly countries might use vegetable cooking oil or butter. Its uses are many and varied. It can be a simple dip for bread or, sprinkled over the most elaborate of dishes, the final touch before serving.

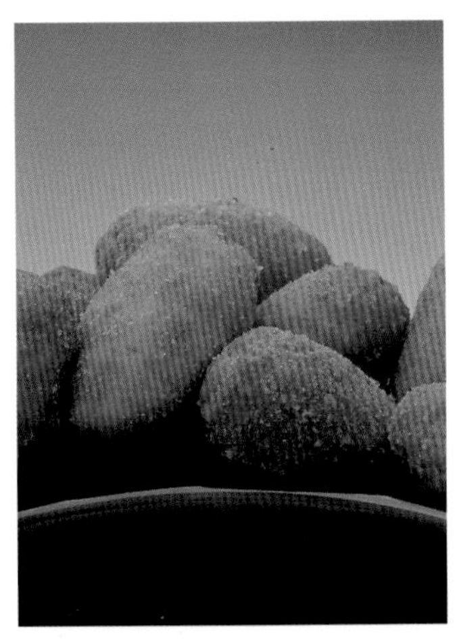

Above: Olive oil has long been used for deep-frying foods such as squid and chicken.

Its most basic use is as a cooking medium to lubricate food for roasting, broiling, and frying. Most people will choose a basic olive oil for this kind of culinary activity but, if you can afford it, try basting your food with extra-virgin oil and taste the difference.

Some believe that olive oil will not stand up to repeated use in deep-frying. In fact, nothing could be farther from the truth. The smoking point of olive oil is the same as corn oil and it can be used in a deep-fat fryer in the same way. However, it does not break down so fast and so can be used more often. It is important to prevent the oil from overheating and to strain it after use.

Olive oil is also used as a condiment to flavor food and as a dressing ingredient for plain dishes such as vegetables, pasta, and salad. This is where the full flavor of extra-virgin oil really comes into its own. Uncooked olive oil is principally used in salads to give a fruity taste to the dressing. Different styles of oil will give different results and it is worth experimenting with a variety of oils, vinegars, and lemon juice to find a range of dressings to use

with different kinds of salad.

Of course, olive oil is the basis of mayonnaise. If you make your own mayonnaise you may find that some extra-virgin olive oils are too strong, but the lighter styles from the northern Mediterranean can be very successful. Alternatively, mix extra-virgin olive oil with ordinary olive oil to get the balance you require.

Around the Mediterranean and in California olive oil is used on its own as a sauce. Try broiling meat or chicken and then drizzle over a little extra-virgin olive oil and some lemon juice. The results are excellent. Olive oil is also added, just before serving, to traditional dishes such as Gazpacho, Ribollita, and Ragout. The heat of the food brings out the flavor of the oil which then blends in with the flavor of the dish.

A fairly recent innovation has been the development of flavor-infused oils. These can add an interesting dimension to your cooking. Pour basil oil into fresh tomato soup, for example, truffle oil onto an omelet or sun-dried tomato oil into a salad dressing. The possibilities are endless.

You can also use olive oil as a short-term preservative for herbed or spiced table olives, roasted vegetables, wild mushrooms or goat's cheese. Some of these delicacies are made by the same estates that produce the oil and you will find a number of them listed in the directory of estate oils later in the book.

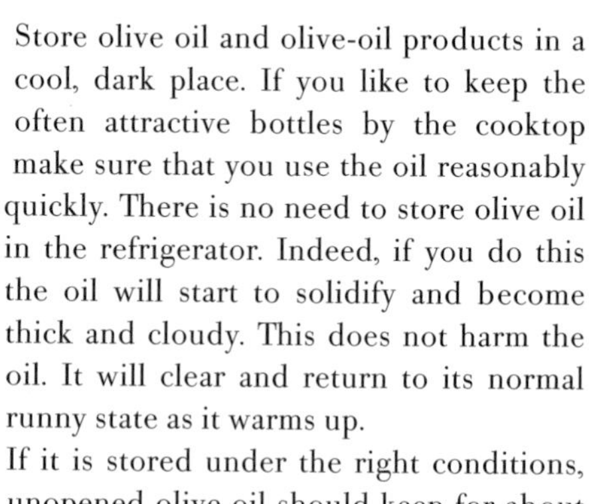

Store olive oil and olive-oil products in a cool, dark place. If you like to keep the often attractive bottles by the cooktop make sure that you use the oil reasonably quickly. There is no need to store olive oil in the refrigerator. Indeed, if you do this the oil will start to solidify and become thick and cloudy. This does not harm the oil. It will clear and return to its normal runny state as it warms up.

If it is stored under the right conditions, unopened olive oil should keep for about a year or so after the harvest. Try to buy it from an outlet which has a reasonable turnover. You do not want to buy oil which is already a year old. Remember that in the northern hemisphere the new season's oil starts to appear in the stores around February or March, depending on how the previous year's supply has moved.

Though only a laboratory test will tell you when an oil is starting to disintegrate or go off, your nose will soon tell you when it is reaching the point that it should not be used. Rancid oil has a very distinctive, unpleasant smell.

Unlike wine, olive oil does not mature in the bottle. It is wise to remember the popular saying, "drink wine old and oil young."

Olive Oil and Health

Olive oil is made up of around 70 percent monounsaturated fatty acids with traces of the antioxidant vitamin E. The rest of the fatty acids are split about 5–10 percent saturated fat and 20–25 percent polyunsaturated fat. It does not contain cholesterol.

Fats and oils are one of the richest sources of energy in the diet. They all produce around 120–125 calories (kilocalories) per tablespoon. So-called light oil refers to oil that is light in flavor not in calories.

The virtues of olive oil in the diet have been increasingly extolled as scientists have discovered the important of monounsaturated fatty acids. Olive oil is made up of around 70 percent of this particular type of fatty acid.

All fats are made up of saturated fatty acids, monounsaturated fatty acids, and polyunsaturated fatty acids and we probably need all three. However, saturated fatty acids are thought to be partly responsible for the high incidence of coronary heart disease in the Western world. Both mono- and polyunsaturated fatty acids are thought to protect against the disease by reducing the cholesterol in the blood.

There are two types of blood cholesterol, HDL (high-density lipoprotein) and LDL (low-density lipoprotein). HDL is regarded as helpful, LDL is not. Polyunsaturated fats have the effect of lowering levels of both the beneficial HDL as well as the LDL. Monounsaturated fat, on the other hand, appears to reduce the LDL levels but leaves the HDL levels unaffected.

Olive oil is relatively easy to digest and helps the assimilation of vitamins and minerals. It has a positive effect on the digestive system, protecting the mucous membranes and stimulating the gall bladder. There is some evidence to suggest that it helps to protect against the formation of gall stones.

Other studies have shown that monounsaturated fatty acids can be helpful for people suffering form non-insulin-dependent diabetes and it is believed they may also have a beneficial role in controlling other potentially harmful conditions.

Information Boxes

Information boxes give the location, estimated yearly production, and an overall 1-, 2-, or 3-star quality rating for each brand. Other at-a-glance information will tell you whether the producers also make wine, and whether visitors are welcome.

The World of Table Olives

Table olives have always been served with drinks and used in cooking in the countries in which they grow, but elsewhere they have not sold so well. However, the pattern is now changing and the range of olives exported is growing rapidly.

Most of the areas which produce olive oil also produce table olives. Other areas specialize in table olives alone so there is a

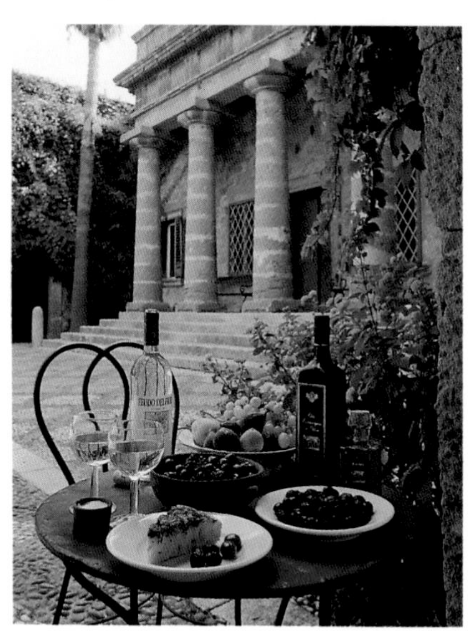

Ravida estate, Agrigento, Sicily.

very wide range of different types from which to choose. They come in all colors from green through pink, red, brown, and purple to black and in all sizes from the tiny Nice olive from Provence to the lusciously large Gordal olive from Spain.

The fresh fruit of the olive tree is very bitter to the taste and is not edible in its raw state. This is because it contains very little sugar and a bitter component called oleuropein. The bitterness must be removed before the olives can be eaten and each region has its own method of curing them. They may be treated with an alkaline lye solution such as sodium hydroxide, they may be fermented in brine or packed in dry salt or they may simply be

cracked and regularly washed with water. Or they may undergo a combination of two or more of these techniques. Each method results in a different texture and flavor.

Above: These are picudo olives, grown in the Cordoba, Granada, and Jaén regions of Spain.

The best table olives are medium-sized (1/8 ounce) or large (approx. 1/6 ounce) though there are exceptions such as the Cailletier or Nice olives. Really black olives are more popular than brown or purple olives because people think these have faded. This could indeed have happened during processing but the lighter color could also mean that the particular variety does not mature to such a dark color or that the olives were picked before being fully mature.

Processing Table Olives

Some varieties are better processed when they are green whereas others are considered to be more appetizing when black. Some processes are traditional and confined to a specific area, others are not. The producer can usually decide from among the following styles:

- Treated with an alkaline solution and fermented in brine
- Treated with an alkaline solution then marinated with herbs and spices
- Treated with an alkaline solution and preserved by heat treatment
- Treated with an alkaline solution and dry-salted
- Fermented in brine without prior treatment
- Partially fermented in brine without prior treatment and heat-treated
- Washed in brine and preserved in brine, vinegar or oil

Spain is the largest of the volume producers in this field and its table olives are famous worldwide. The best come from Andalusia in the south from Seville, Córdoba, Málaga, Jaén, and Extremadura. The majority of the olives grown are processed when they are still green and stuffed with a variety of fillings, including pimento, garlic, and almonds.

The typical Sevillian process involves treating the olives with an alkaline solution to remove the bitter elements then washing the fruit with successive water baths to remove the lye and any other bitter residues. The olives are then fermented in brine. Once the process is complete the olives will keep for comparatively long periods of time.

There is a tendency in Spain to preserve and pack cheaper olives in a chemical salty solution and these are best avoided. Look out instead for specific varieties, such as the Gordal or "Queen" olive and Manzanillos, which have been treated with more care.

Italy is better known for its olive oils than for its table olives. However, most of the producing areas process table olives as well as olive oil but most of them are consumed in Italy. A variety of different methods are used to cure the olives. Italians cannot get enough olives and Italy is a net importer.

About 20 percent of the annual Greek olive crop goes to produce table olives. They are mainly harvested when they are fully ripe. Typical processing methods would include fermenting the olives in airtight containers or large vats. Color tends to be lost from the fruit during fermentation and this may be replaced by aerating the olives for a few days or by adding ferrous gluconate to make them darker.

French table olives are known for their very fresh flavors. One of the most popular types is "olives cassées," broken olives. The freshly-picked olives are given a hard tap to split them open. They are then treated with caustic soda, washed frequently and marinated in herb-flavored brine. Every producer has his own secret recipe and the broken flesh absorbs the flavors particularly well.

The Californians have developed a technique for picking unripe or partially ripe olives and allowing them to darken as they are processed. The olives are treated with an alkaline solution and then aerated. This sequence is repeated until the olives have softened and the color has darkened. They are then heat-treated and canned.

Common Varieties of Table Olive

Aragon comes from northern Spain. It has a jet-black color which makes it extremely popular. It is oval in shape, smooth, and glossy.
Arbequina olives (also used for oil) come from northern Spain. They are quite small and an unusual, dark brownish-green color. They are often sold in oil flavored with rosemary and thyme.
Cailletier olives (also used for oil) come from the coastal mountains behind Nice in Provence. Also known as Nice or Niçoise, these tiny olives have a reddish tint.
Conservolea olives come from Greece. They are picked when they are fully ripe and fermented in brine in airtight containers. They have a slightly bitter aroma and taste.
Gordals or "Queen's Olives" come from the area around Seville in southern Spain. They are large and fleshy, with a large pit. They may be processed green or black.
Hojiblanca olives (also used for oil) come from Jaén and Córdoba in southern Spain. They are more fibrous than many olives and not as tasty as some of the other Spanish olives.
Kalamata olives come from Kalamata in the Peloponnese in Greece. They are quite large and juicy and have a good black color. They have a low bitterness level and so are easy to cure. Incisions are made in the olives which are then washed in water or brine. This is followed by one or two days soaking in wine vinegar before being packed in oil, brine or with herbs or pieces of lemon.
Lucque olives come from Languedoc in southwest France. They are a superb green table olive, with a long, thin pointed pit. They may be prepared "à la Picholine" (see below) or as "olives cassées."
Manzanillo olives come from Andalusia in southern Spain. They are the most delicate and smooth-tasting of all the Spanish green olives. They come from the Manzanillo olive tree which is also widespread in California.
Megaritki olives come from Greece. They are picked when they are overripe and processed by washing and stacking in dry salt. The olives shrivel to the size of a raisin. They are salty but not at all bitter.
Mission olives come from California. They are medium-sized and are usually treated with lye and air and then canned.
Picholine olives come from Languedoc and Provence in southern France. They are lovely green olives that are long and thin. They take their name from the curing process which was invented by a

M. Picholine. He developed a method of curing green olives with fine wood ash from the oak trees which grow on the local scrub.
Salonenque olives come from the Hérault and Aude regions of central southern France. They are green olives, usually prepared as "*olive cassées*."
Taggiasca olives come from Liguria in northern Italy. They are small and black and very similar to the French Cailletier.
Tanche olives (also used for oil) come from Nyons in northern Provence. They were the first table olives to be given an A.C. guarantee of origin. They are black in color. They are picked when they are slightly overripe with fine wrinkles caused by the first frosts.

Serving Table Olives

Cured olives can be packed for sale in brine, dry salt or olive oil. After processing some olives are stoned and stuffed. Others are flavored with herbs and spices. Once cured some table olives are minced to form olive pastes or to be used in products such as the French Tapénade or Anchoïade.

There are no rules about when and how to serve table olives. Each region serves its own style of table olive with a glass of wine, an aperitif or a stronger drink. No selection of Spanish tapas is complete without two or three kinds of olive and the same goes for a Greek meze. A small plateful will also appear as an appetizer at the beginning of a meal in Italy and Provence.

If you like a particular variety of olive which is not usually flavored with herbs or spices you can make up your own mixture of flavors and toss the olives in it or steep herbs, olives and other ingredients in olive oil. Serve stuffed olives on cocktail sticks on their own or with fried croutons, cubes of cheese or ham.

Use olives as a garnish on dishes like Paella or Jambalaya. Add to green salads in the Greek manner or to orange salads in the Spanish style. Use them to add interest to casseroles, stewed vegetable dishes, and pasta. The possibilities are endless.

DIRECTORY OF OLIVE OILS

2

Abbo

Choose this gentle extra-virgin olive oil to make mayonnaise, says Gianpaolo Abbo, nephew of the company's founder and now chief oil taster. He is right, the result is fragrantly delicious.

Secondo Abbo set up his olive mill in Ventimiglia at the French end of the Italian Riviera in 1893. He was a familiar sight riding his heavy bicycle – it weighed more than 22 pounds – over the steep mountain passes to drum up business.

Since then the company has grown considerably and is now a member of the prestigious Mastri Oleari, the Oil Masters Corporation. It has plants in three centers but the local Taggiasca and Ogliarola olives are still pressed in Ventimiglia. If you want to see the original premises visit Abbo in midwinter when the mill is in full swing. At other times of the year you will have to make do with the offices and high-tech packaging plant in Saluzzo in Piedmont, 60 miles to the north.

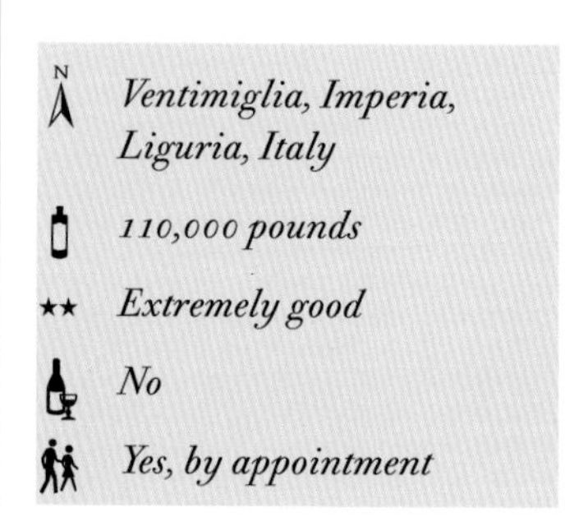

Abbo

The olives for the Abbo oil come from different groves in the steep valleys which run inland from the coast in this part of Liguria. They are picked by hand when they are very ripe and this gives the oil its particularly sweet and mild flavor. The Abbo mill is often still in operation when the others have closed down.

The oil has a delicate grassy aroma with a hint of apples. The flavor is pleasingly nutty with light to medium pepper. As well as making good mayonnaise, the oil is used locally to bake cakes and cookies.

Bruschetta with Anchovies

Drain the oil from a 2-oz can of anchovies in olive oil and mince the anchovies in a bowl or food processor. Add four minced cloves of garlic and the juice of half a lemon. Very gradually add ⅔ cup extra-virgin olive oil, beating with a fork or running the processor. The texture should be like mayonnaise. Spread onto toasted country bread and dot with halved, pitted black olives. This can also be served as a dip.

Product Box

Extra-virgin olive oil

Olives in brine
Pitted olives in olive oil

Pesto sauce, artichoke paste in olive oil, and arugula cream in olive oil

AFFIORATO MANCIANTI

This wonderful extra-virgin olive oil is the "first run" of the milled olives. No press is used, the oil runs freely from the paste.

Affiorato Mancianti is a member of the Mastri Oleari. The olives include Moraiolo, Frantoio, and Raggiale. The olives are picked by hand and stone-milled for slightly longer than usual. The naturally-surfacing oil which emerges from the paste after milling and before pressing is packed as Affiorato.

This is a very sweet and elegant oil, leaving a deliciously lingering taste in the mouth. Both the aroma and the taste are full of fresh olives, salad leaves, and lemons. The oil is very smooth, with a touch of peppery bitterness before the sweet aftertaste.

PRODUCT BOX

Affiorato Mancianti extra-virgin olive oil
San Feliciano extra-virgin olive oil
Monte del Lago extra-virgin olive oil

Black olive paste: with and without chili

San Feliciano sul Trasimeno, Perugia, Umbria, Italy

792 U.S. gallons Affiorato and 2,000 U.S. gallons of other extra-virgin oils

★★★ *Excellent*

No

Yes, by appointment

A L'Olivier

Belief in the health benefits of olive oil is not a twentieth-century phenomenon. This company was founded on just such a conviction way back in 1822.

A l'Olivier was created by M. Popelin, a Parisian chemist, who gradually extended his activities to become *the* French specialist in quality oils. In 1978 the company was bought by Jean-Claude Blanvillain, another oils expert, who moved the company to Provence. It is now run by M. Blanvillain and his two sons from premises outside Nice.

The company buys good-quality extra-virgin olive oils and blends them to produce a consistent flavor every year. The oils are simple and straightforward with a good, lemony olive flavor. They are smooth and not very peppery and are generally good for all-round cooking.

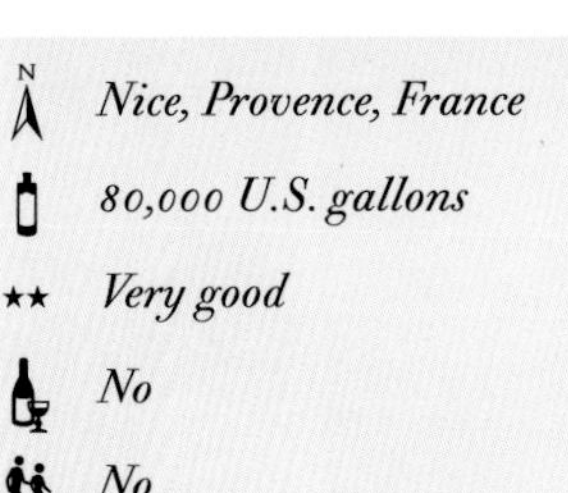

Nice, Provence, France

80,000 U.S. gallons

Very good

No

No

Above: The A L'Olivier Store in 1931.

Product Box

Extra-virgin olive oil "Mère Goutte"

Extra-virgin olive oil "Mère Goutte" Fruty

Olive oil with infused fresh basil

Provence table olives

Black and green olive tapénade

Red peppers in olive oil

Moulin à Huile

Alziari

This oil mill in fashionable Nice on the French Riviera has been in the Alziari family since 1878. It specializes in processing the tiny Caillette or Cailletier olives.

These olives, also known as Nice olives, are among the smallest and it is only their highly concentrated, fine flavor which makes them worth bothering with. They are used at the Alziari mill to make both oil and table olives.

The Cailletier olive groves are situated high in the granite mountains behind Nice. When picking starts the groves are bright with the red nets. These are spread on the ground to catch the olives as they are taken off the trees with air-operated rakes. The olives are then taken to be pressed at the traditional mill.

Alziari extra-virgin olive oil has a really fresh lemony aroma with apples and a delicate flavor of lemons, apples, and a little cut grass. The pepper is very light.

Product Box

Extra-virgin olive oil

Olives de Nice
Green olives Picholine

Olive pastes, tapénade, and anchoïade

N *Bd. de la Madeleine, Nice, France*

Not known

★★★ *First-class*

No

Yes

Antinori

Laudemio and Peppoli

The Marchesie Pietro Antinori is famous for his Tignanello "super-Tuscan" wine and for his Chianti but his estates also produce first-class extra-virgin olive oil.

The Antinori history spans 26 generations and stretches back over 600 years of both traditional and innovative farming. The family owns the fifteenth-century Palazzo Antinori in the center of Florence as well as a number of farm estates in the Chianti region where the oils and wines are made.

The quantity of olive oil produced on the Antinori estates has grown considerably over recent years, though the 1985 frosts were something of a setback at the time. However, they did inspire research into methods of cloning for frost resistance and into new methods of cultivation. These are now being put into operation and it is hoped that the trees will survive cold snaps rather better in the future and allow more constant production.

Today there are 326 acres of olive groves on the estate, planted with Frantoio, Leccino, and Moraiolo olives. On the Peppoli farm, close to the town of San Casciano Val di Pesa, there are 5,500 olive trees, some of which survived the frosts and date back

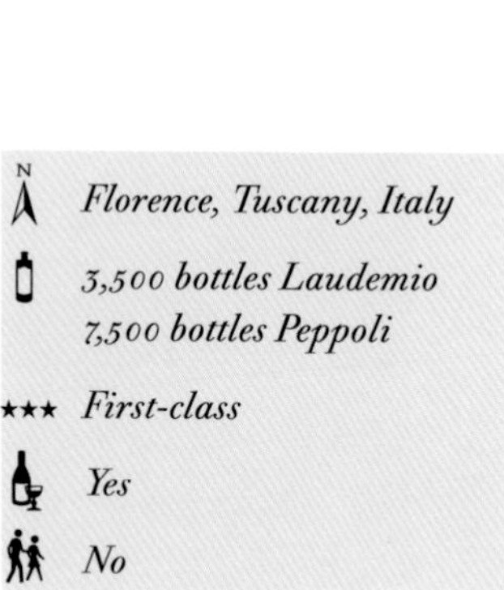

Above: Overlooking olive trees towards the Peppoli estate.

several hundred years. Olives are also grown on the Badia Passignano and Santa Cristina farms.

Olives from all three farms go into the blend for Laudemio Tenuta Marchese Antinori oil. They are crushed in the traditional manner at the picturesque stone oil mill at Santa Cristina. The Peppoli oil is made only from olives grown on the Peppoli farm. Here mechanical shakers are used to encourage the olives to fall into the nets and the olives are processed by the cold centrifugal method.

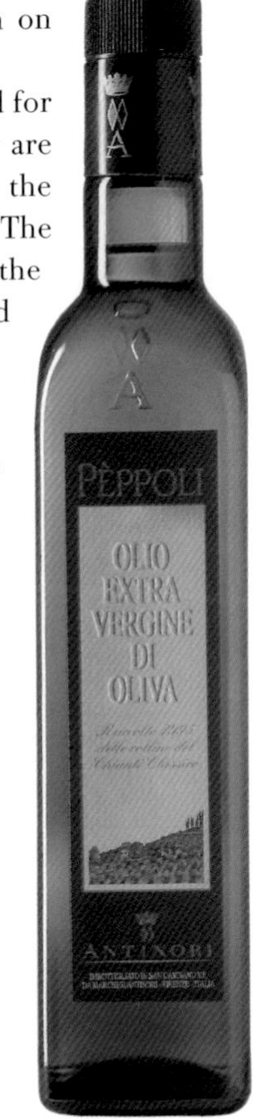

The Laudemio oil has a fragrant aroma of freshly-cut grass, arugula salad leaves, and sorrel. The flavor is rich with fresh olives and nuts. It has a fiery kick to it but a sweet aftertaste.

Peppoli extra-virgin olive oil offers an aromatically herbaceous aroma of cut salad leaves and grass. The flavor is a complex mixture of sorrel, watercress, and arugula with lots of pepper and some toasted nuts on the aftertaste. It is a smoothly elegant oil.

Above: Piero Antinori with his wife, Francesca.

Antinori Panzanella Salad

The recipe for this unusual salad comes from the Cantina Antinori in Florence where you can sample both the Antinori oils and wines. Place 4 cups coarsely cubed day-old firm Italian bread on a baking sheet and drizzle over 2 tablespoons extra-virgin olive oil. Bake in a hot oven for about 8–10 minutes, turning occasionally until crisp and lightly browned. Allow to cool and mix with 4 large chopped tomatoes, a small sliced red onion, a large bunch of coarsely chopped basil, and 4 or 5 leaves of Romaine lettuce torn into pieces. Now whisk 2 tablespoons of red-wine vinegar with six tablespoons of extra-virgin olive oil. Season and pour over the salad. Mix gently and allow to stand for 10–15 minutes before serving.

Pietro Isnardi

Ardoino and Isnardi

Vallaurea and Fructus

Two old-established Ligurian family businesses have now joined forces to produce and sell a range of extra-virgin olive oils with a prestigious reputation.

The Ardoino family started making olive oil at Pontedassio near Imperia around 1870. They hit on the idea of delivering the oil by boat to customers in the ports further along the Ligurian coast. This was followed by a period between the World Wars when they sold their oil exclusively to America.

Then, after World War II, Nanni Ardoino set about making particularly delicate oils for discerning customers. Today he continues to follow the selection of the Ardoino oils but has handed the business over to Pietro Isnardi whose family have also been dedicated oil producers since the turn of the century.

Now the company has both the original mills and new premises on the outskirts of Imperia. The old presses are linked to modern Sinolea machines which cut the olive paste with 5,000 knives extracting the oil by surface tension.

The Ardoino Vallaurea gets its name from the high valley of Oneglia. This is where the Taggiasca olives which go into this oil are grown. The fruit is picked when it is ripe and there is an unusually long harvest which can last from November until May.

The pickers use poles to knock the olives into nets spread on the ground. They then go to the mill for the oil to be extracted. Ardoino oil is never filtered. It has a very delicate aroma of lemons and almonds, with almonds predominating. It is not at all aggressive and there is very little pepper.

Ardoino Biancardo is probably the rarest oil in the world. It is produced in very small quantities when the maturation of the Taggiasca olive extends into April and May. The taste is said to be sweet with an aftertaste like delicate velvet.

Ardoino Fructus is a blend of oils made from Taggiasca and a variety of other Mediterranean olives. The flavor is lightly fruity with a touch of grassiness and a peppery aftertaste.

Product Box

Ardoino Vallaurea
extra-virgin olive oil
Ardoino Fructus
extra-virgin olive oil
Ardoino Biancardo
extra-virgin olive oil
Isnardi Special Selection
extra-virgin olive oil

Table olives, pitted and unpitted, in brine

Black olive paste

Pesto sauce

Anchovies in olive oil

Vegetables and vegetable pastes in olive oil

Tenuta di Castello di

ARGIANO

This estate near Sant'Angelo-in-Colle in southern Tuscany forms part of the portfolio of a multinational drinks corporation but it is managed locally from Montalcino.

There are 20 acres of olive groves and 24 60 acres of vineyards, surrounding the old manor house with its splendid wine cellars. The groves are planted with Frantoio, Leccino, and Pendolino, with Frantoio as the predominant variety.

The olives are hand-picked and processed in a Sinolea machine which uses knives to cut the paste and relies on surface tension to extract the oil.

Argiano extra-virgin olive oil has an interesting aroma of olives, cut grass, and toasted nuts. The flavor is particularly unusual as it seems to change from bitter grassiness to rich nuts, chocolate, and even coffee. There is plenty of pepper, softening to an increasing smoothness.

Sant'Angelo-in-Colle, Montalcino, Tuscany, Italy

800 U.S. gallons

★★★ *First-class*

Yes

Yes, appointment only

Fattoria Michelangelo

Aristeo

According to local legend Aristeo, the god of the earth and its fruits, introduced the olive to the Bitonto region of southern Italy. Fattoria Michelangelo have named their extra-virgin and flavored oils after him.

Only Coratina and Ogliarola olives are used for this oil. they are grown in groves around Terlizzi which is situated inland from Bari, northwest of Bironto Bari in Apulia. The olives are hand-picked and the oil is extracted by continuous cold centrifugal process. It is unfiltered.

The oil, which has gained the company entrance into the Mastri Oleari, has a pleasantly warm aroma of apples and lemons with some nutty undertones. The flavor is fresh and full of almonds with a touch of celery and herbs. There is just a little pepper on the finish. It is a versatile oil which can be used in most culinary applications.

Product Box

Aristeo extra-virgin olive oil

Aristeo flavored olive oils

Capurso, Bari, Apulia, Italy

Not known

★★ *Extremely good*

No

No

Athena

This is one of the few Greek extra-virgin olive oils which claims to be pressed from 100 percent Kalamata olives. The flavor is not at all sweet and is quite distinctive.

The olives for this oil come from groves varying from 50 to 5,000 trees which are farmed as small cooperatives in the Kalamata and Messinia regions of the southern Peloponnese. Because of the altitude few or no pesticides are needed.

The harvest starts before the olives are fully ripe in early to mid October on irrigated groves and mid November on unirrigated groves. Both traditional and centrifugal methods of processing are used.

This oil has a lightly nutty aroma with some grassy hay. The taste is quite bitter but rather attractive with the sour flavor of sorrel leaves. It is an unusual oil which cannot be ignored.

Marinade for Swordfish or Tuna

Marinate swordfish or tuna steaks in plenty of extra-virgin olive oil mixed with 2 tablespoons each of freshly-squeezed lime and lemon juice, garlic, chopped chives, salt, and pepper. After an hour or two remove the steaks from the marinade and broil or fry.

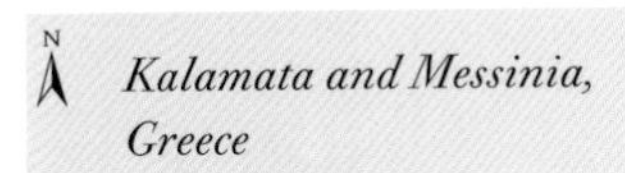

Kalamata and Messinia, Greece

Not known

★★ *Very good*

No

No

Azienda Agricola
AVIGNONESI

Three estates under the management of one family contribute to this sweetly nutty extra-virgin olive oil from southern Tuscany.

The Falvo brothers took over Avignonesi in 1974 and have spent the intervening years building up both the wines and the olive oil. The Le Capessine estate at Valiano near Montepulciano lies at the heart of the company. Surrounded by 100-year-old oak trees the buildings, including the *frantoio* or oil-mill, have been restored. There are a total of five acres of olive groves here.

I Pogetti, situated near Argiano, and La Selva at Cignano in Cortona add 12 further acres of olive groves. Correggiolo is the dominant olive variety, providing 80 percent of the oil. The rest is made up of both Moraiolo (15 percent) and Leccino (five percent).

The fruit is picked by hand in November and then stoneground and hydraulically pressed. It is then stored in the dark in the traditional terracotta bowls of the region for three to four months to clear.

This olive oil has a light lemony apple aroma with a touch of almonds. The overall impression is gently sweet. It tastes equally smooth and sweet with a rich nutty flavor and a little chocolate. The pepper is very light.

N *Valiano, Montepulciano, Siena, Tuscany, Italy*

1,000 U.S. gallons

★★★ *First-class*

Yes

Yes, by appointment

Badia a Coltibuono

Badia a Coltibuono and Badia Albereto

Situated in the heart of the Chianti region this ancient estate with its old monastery is the center of a family business embracing olive oil, wine, and gastronomy.

Pietro Stucchi Prinetti is the principal inspiration behind the philosophy of perfection that pervades the estate, but all the family are involved in the work. One of his sons is the oenologist and manages the estate and the other runs a restaurant in the grounds serving traditional Tuscan food. His daughter is in charge of marketing and public relations.

Lorenza de Medici, the well-known cookery writer, is Pietro Stucchi Prinetti's wife. She has founded a cookery school that attracts many students to the monastery to learn about the best of Italian gastronomy.

Above: The beautiful buildings of the old monastery.

Gaiole-in-Chianti

3,300 U.S. gallons

★★★ *First-class*

Yes

Only by appointment in groups

Above: Pietor Stucchi Prinetti and his family.

Badia a Coltibuono is set among the Chianti hills near Gaiole-in-Chianti about halfway between Florence and Siena. The olive groves are planted with 70 percent Frantoio, 20 percent Leccino, and ten percent Pendolino. The harvest is picked early, following the usual practice in this part of Tuscany. The olives are hand-picked and pressed in the traditional manner. The oil is then filtered through cotton wool.

Badia a Coltibuono extra-virgin olive oil has a sweetly fruity aroma of cut grass and grated apples. The flavor is equally fruity with a touch of toasted nuts which is very attractive and which increases as the oil ages. There is plenty of pepper but it is not at all aggressive.

The Badia Albereto extra-virgin olive oil has a sharper aroma of cut grass and sweet-sour salad leaves. The flavor is distinctively bitter with a real peppery kick, but the aftertaste is sweetly grassy. A really interesting oil.

Pinzimonio

This is the Italian inspiration for the French crudités, or so Tuscans will tell you! Gather as interesting a collection of raw vegetables as you can find. This may include fennel, radishes, radicchio, tomatoes, artichoke hearts, red bell peppers, cucumber, celery, and carrots. Clean and cut into strips or chunks and serve with a large bowl of well-flavored extra-virgin olive oil and salt. Give each person a plate and a small dish. The latter allows everyone to mix their own olive oil dip, adding salt and pepper and sometimes vinegar to taste.

Abasa

BAENA

Campoliva

Every bottle and every can of Baena extra-virgin olive oil which leaves Abasa carries a guarantee of origin.

Baena is situated in the province of Córdoba. It has 79,000 acres of olive groves, but most of the growers send their oil to large cooperatives or companies such as Abasa for marketing. The main olive varieties are Picual, Picudo, and Hojiblanca. The groves and mills are registered by the Quality Control Council. Once the oil has been pressed it is sent for analysis and tasting. Only then can it be labeled as D.O.C. oil.

Baena oil has a fresh lemony, grassy aroma with a hint of melon fragrance. The taste is pleasantly sweet with a little pepper to the aftertaste. It is a versatile oil with many uses.

PRODUCT BOX

Baena D.O.C. extra-virgin olive oil
Campoliva extra-virgin olive oil
Olivaena extra-virgin olive oil
Guadoliva extra-virgin olive oil

Córdoba, Andalusia, Spain

3,520,000 pounds

★★ *Extremely good*

No

No

Fattoria

Baggiolino

Laudemio and Baggiolo

These two typically green and peppery Tuscan extra-virgin olive oils owe their quality to the efforts of Ellen Fantoni Sellon and her two daughters who run the estate with the help of farm manager Andrea Poli.

The Fantoni Sellon family bought land from the local Casignano estate in the 1950s and have built it up into this successful farm.

There are about 39 acres of olive groves planted around the vineyards. The estate, which nestles in the hills near Florence, also includes 165 acres of woodland which almost entirely surrounds the cultivated areas. This creates an ideal microclimate which is dry and temperate.

There is a saying in Italian that to have good oil you need the three S's, "secco, sasso e sole" which means dry climate, good soil, and sunshine. The soil here is rich in clay and marl and Ellen Fantoni Sellon believes

La Romola, Florence, Tuscany, Italy

13,000 pounds

★★★ *Excellent*

Yes

Yes, by appointment only

that this contributes to the distinctive taste of her oils.

Both the olive oils are made from a blend of 65 percent Frantoio, 15 percent Moraiolo, and ten percent Leccino and ten percent between Pendolino and Morchiaio. The fruit is picked between 1 November and 15 December. The oil is extracted by the cold centrifugal system.

The Baggiolo extra-virgin oil is very green with an aroma of salad leaves and apples with sorrel. The flavor is grassy with cut salad leaves and some toasted nut. It packs quite a peppery punch but the aftertaste is elegantly smooth.

The Laudemio Baggiolino extra-virgin oil is even more grassy with almonds, tart sorrel, and a mixture of other wild salad leaves. The flavor is equally distinctive with fiery pepper and a sweet finish. The pepper in this oil tends to soften after a while.

White Beans in Olive Oil

Cook some white beans such as cannellini beans or black-eyed beans until they are tender. Drain well and toss with freshly chopped parsley and a little garlic if you like it. Season and carefully toss together. Dress with plenty of well-flavored extra-virgin olive oil and serve with broiled or roasted meat.

Azienda di

Bartolini

This Umbrian estate has been in the Bartolini family since 1850 and their untiring efforts to increase the quality of their produce have resulted in a prize-winning extra-virgin olive oil

Situated near Valnerina in the province of Terni the olive groves are planted mainly with Moraiolo, with Frantoio and Leccino as the back-up olives. The fruit is hand-picked and the oil is extracted by the cold centrifugal system.

Bartolini extra-virgin olive oil has a rich nutty aroma with spices, eggs, and lemons. The flavor is equally complex with toasted nuts and coffee tones moving to bitter leaves and warm pepper. The aftertaste is very smooth and nutty. An interesting oil for all kinds of culinary uses.

Bruschetta with Tomatoes

Toast thick slices of Italian country bread. Top with sliced tomatoes and broil. Sprinkle with salt and drizzle extra-virgin olive oil over the top. Top with some torn basil leaves.

Valnerina, Terni, Umbria, Italy

2–3 tons

★★★ *First-class*

No

Yes

Oleificio
Borelli

Founded in 1984, this relatively new company has established itself as a major producer with branches in the U.S., Canada, and France.

The range of oils available from Borelli includes specialty Ligurian and Tuscan oils as well as branded Borelli extra-virgin oil.

The Ligurian extra-virgin olive oil is pressed from 100 percent Taggiasca olives grown on farms around Borgomaro in the Imperia Valley.

Borelli's Taggiasca extra-virgin olive oil has a light apple aroma and tastes quite sweet and nutty with some pepper. It is a straightforward but attractive oil.

The Tuscan extra-virgin olive oil is pressed from a blend of 65 percent Frantoio, 25 percent Leccino, and ten percent Moraiolo olives grown on farms around Pieve a Elici, near Lucca. The fruit is harvested in December and January and is pressed in the traditional manner.

The Borelli brand extra-virgin olive oil has a sweet, lemony aroma with a more herby and grassy flavor which is at its best when it is very fresh. The oil is quite rich and eggy without too much pepper. Occasionally the blend can be too sweet.

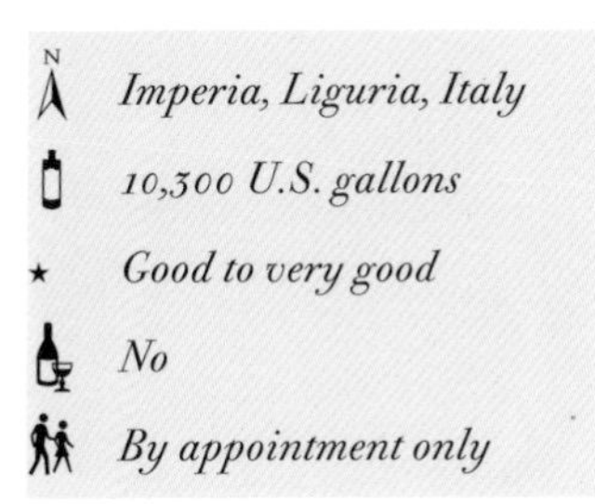

Imperia, Liguria, Italy

10,300 U.S. gallons

Good to very good

No

By appointment only

Aceites

Borges

Though it is now one of Spain's largest producers of olive oil, this firm is still 100 percent Catalan and 100 percent family owned.

The Borges company was established in 1896 by Antonio Pont Pont. Three generations later the company still deals in products based on the two items with which its founder started – olive oil and almonds.

The olives for the Borges oils come from numerous groves in the gentle hills and valleys of Catalonia in northeastern Spain. They produce an extra-virgin olive oil which has a quite lush and intense aroma of melons and passion fruit with eggy overtones. The flavor is sweet and easy, with less fruit than you might expect from the aroma. The taste moves on to quite a bitter finish with heavy pepper.

This is a very definite oil which goes with well-flavored foods. Serve with gazpacho, arugula and watercress salad and broiled meats.

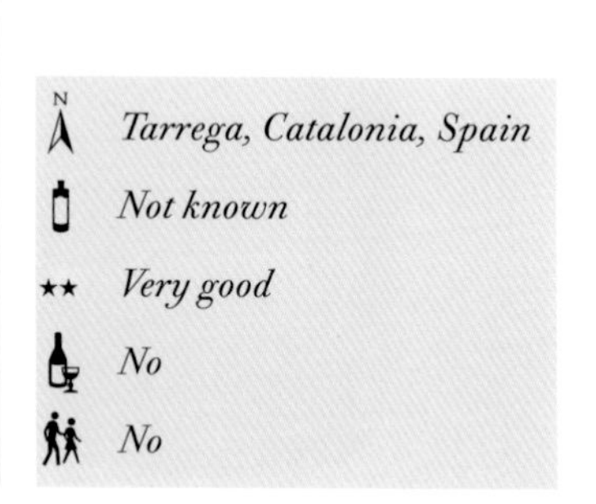

Azienda Agricola Fratelli

BOTTARELLI

Three generations of experience go into the production of this sweetly fruity extra-virgin olive oil from the western shores of Lake Garda in Italy.

The Bottarelli family grow their olives and tend the vines in the hills around Polpenazze del Garda in Lombardy. Signor Bottarelli senior makes the wines and his two sons look after the olives and market the produce of the farm.

The principal olive varieties are Easaliva and Leccino. The blend produces a fresh and fruity olive oil with apple and lemon tones on both the aroma and the taste. It is a pleasant and undemanding oil which can mature to a more nutty and earthy flavor.

The oil is not very peppery and can be used for almost all culinary purposes. It is particularly good served warm drizzled over fried fillets of white fish or slices of fried polenta topped with an egg.

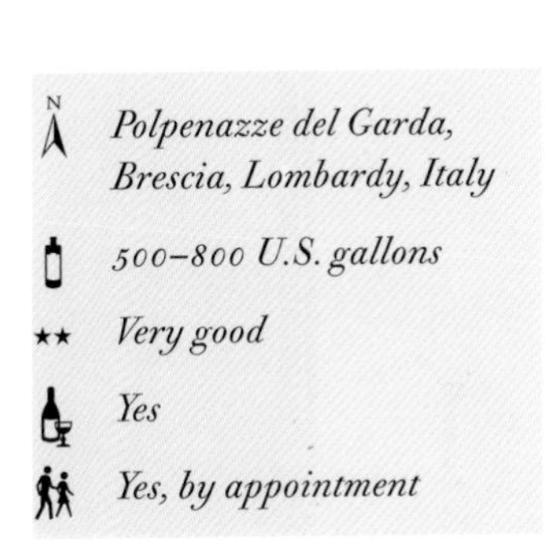

Coop. Agricola Brisighellese

Brisighella

The olive groves surrounding the beautiful and historic town of Brisighella in the Appennines supply the olives for this oil.

The cooperative at Brisighella was formed in 1972 to preserve the high quality of the local produce. From 1975 each bottle leaving the mill has had a serial number and a warranty so if anything does go wrong the fault can be traced back. They are now members of the prestigious Mastri Oleari.

Brisighella lies in the Lamone valley on the panoramic road from Faenza to Florence on the Emilia Romagna or eastern side of the mountains. It has plenty of sunshine and is protected from the cold northerly winds and the olive trees flourish. Farmers grow an ancient local variety of olive called Nostrana which literally means "local" in Italian.

Picking starts at the end of October and lasts until December. The olives are hand picked then pressed traditionally. The oil is left to decant naturally until January when the best oil is selected to carry the Brisighella name.

Brisighella extra-virgin olive oil is attractively complex with an extremely fresh aroma of grass and apples and a hint of nutty chocolate. The flavor is warm and nutty with plenty of fruit and strong pepper. The aftertaste is soft and sweet.

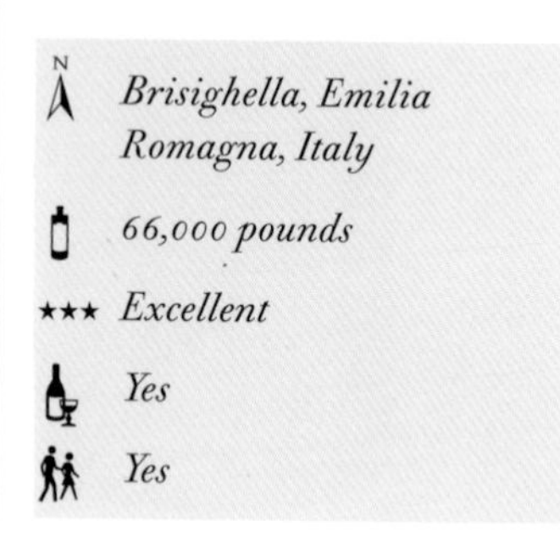

Olive Oil and Land Company

CALAVERAS

Ed Rich, President of Calaveras, is one of the new band of Californian olive oil pioneers but his own groves will not come on stream for another year or two.

In the meantime Ed Rich buys the best local olives he can find for his range of olive oils. They are pressed and bottled at the Golden Eagle olive estate in the Portersville area of California.

This year they have pressed and bottled a number of different Calaveras extra-virgin olive oils. The most interesting is the Motherlode Blend. It is pressed from Mission, Manzanillo, Picholine, and Navidillo olives grown in two small groves near to Copperopolis. The aroma is very fresh with apple skins and a touch of grass. The flavor is fruity, with almonds and medium pepper. It fades a little on the aftertaste but is nevertheless very good.

The Calaveras company itself is based on an old cattle ranch near Copperopolis, California. The new groves are planted in the Sierra Foothills in an area which used to produce a large quantity of table olives. However, Ed Rich is convinced that the way ahead is not to replant the old Californian olive varieties but to use Italian, Greek, and French varieties. He has already planted 500 trees and plans to add more. To date they include Frantoio, Leccino, Kalamata, and San Felicia varieties.

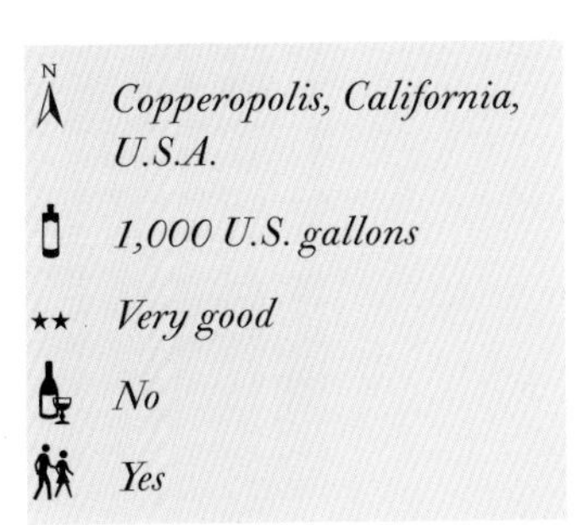

Tenuta di CAPEZZANA

Count Bonacossi and the members of his family run this beautiful estate and villa in the hills of Carmignano. World-famous wines and extra-virgin olive oil are the result of their labors.

The oldest written document on Capezzana dates back to the time of Charlemagne to a contract in A.D. 806 which states that the estate was rented out with vineyards and olive orchards. Various families owned the estate until the beginning of this century when it was inherited by the Bonacossi family.

The countryside is steeply hilly with an attractive mix of olive groves, vineyards, trees, and small fields. All the olives for the Capezzana oil are grown on the estate. This has not always been easy to achieve as the area is prone to frost which can severely damage and even kill some of the trees.

The oil is blended from 45 percent Moraiolo, 30 percent Frantoio, 15 percent Pendolino, and ten percent Leccino, though the mix might vary a little from year to year. The olives are hand-picked by brushing onto nets and then stone-ground. This is followed by the centrifugal separation of oil and solids. This half-and-half traditional and modern method is quite popular in Tuscany.

The result is a light and fresh aroma of apples, grass, and leaves. The taste is smoothly sweet with a flavor reminiscent of salad leaves. There is plenty of pepper but the aftertaste is as sweet and smooth as the start.

N *Carmignano, Florence, Tuscany, Italy*

Not known

★★★ *Excellent*

Yes

Yes, by appointment

Carapelli

Oro Verde

Carapelli is the leading olive oil brand in Italy. The company produces some 12 million U.S. gallons of oil every year at a new factory set in the Tuscan hills a few miles south of Florence.

The company was set up by Constantino Carapelli in 1893 and sold both grain and oil. Over the years the company has come to specialize in oil, though it also produces some other olive products.

Carapelli Oro Verde extra-virgin olive oil is a straightforward oil with a grassy aroma and flavor. A touch of tart sorrel leaves adds interest to the aftertaste. It is quite peppery.

Use for salad dressings made with red-wine vinegar and mustard, marinades for spareribs and barbecued chicken or pot roasts of beef and lamb.

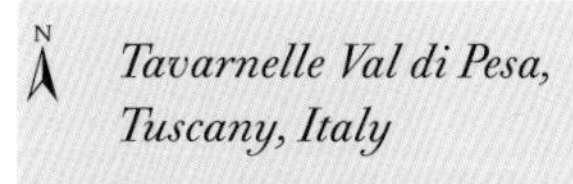

Tavarnelle Val di Pesa, Tuscany, Italy

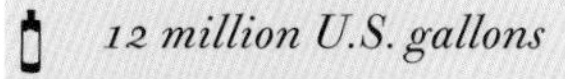

12 million U.S. gallons

Good

Yes

Yes, by appointment

Aceites

CARBONELL

Started by Antonio Carbonell in 1866, this Spanish oil giant once had a contract to supply the British Admiralty ships with olive oil. Today it sells its range of oils all over the world.

Carbonell produces consistently good-quality extra-virgin olive oil. The company blend oils from Córdoba and Jaén in Andalusia and the style is typical of the region. Arbequina, Hojiblanca, Picual, and Picudo are the local olive varieties. They are hand-picked around the middle of December prior to cold centrifugal processing.

Carbonell extra-virgin oil has a sweet and fruity aroma offering a mix of lemons, melons, apples, and grass. The flavor is not quite so intense but it is still warm and fruity with a hint of lychees. Light to medium pepper makes this an easy oil to use. Look for the new season's oil in the spring when it really is at its very best.

PRODUCT BOX

Extra-virgin olive oil
Special selection extra-virgin olive oil

Black and green olives

Córdoba, Andalusia, Spain

260,000 pounds

★★ *Extremely good*

No

By appointment only

Stefano

Caroli

Monte del Duca and Monte Trazzonara

This successful Apulian oil company offers a wealth of choice in its four extra-virgin olive oils. Each oil has its own distinctive style.

Based on Martina Franca in the center of the hilly country which makes up the heel of Italy, this company presses its oils from olives grown on their own 44-acre olive grove. These are supplemented with other olives grown in the surrounding areas. The actual mix of olives which goes into each oil is regarded as a company secret.

The olives are harvested by hand by shaking into nets and most of the picking is over by the end of December. The company's Caroli extra-virgin oil is produced by the continuous centrifugal process. This attractively aggressive oil has a good fruity aroma reminiscent of freshly grated apples but it tastes much more grassy and nutty. It has a strong peppery kick.

C/da Trazzonara, Martina Franca, Italy

6,700 U.S. gallons from olives grown on the company's own estate plus 130,000 U.S. gallons from purchased olives

★★ *Extremely good*

No

Yes

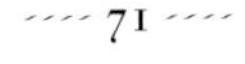

Caroli Monte del Duca imitates the pleasing grass and apple flavors of the standard Caroli oil but it is much sweeter and more elegant. The flavor is full and fruity but not aggressive and the peppery aftertaste is lighter. This oil is made by the traditional method.

So too are Masseria Monte Trazzonara and Antica Fattoria Caroli. The Monte Trazzonara is an interesting oil. It offers light nuts, apples, and tomatoes on the aroma but it is deeply nutty, almost chocolaty, in the mouth with an astringent and lightly peppery kick reminiscent of arugula or watercress. Antica Fattoria Caroli is also a darker, nuttier oil but it does not have the finesse of the Monte Trazzonara.

Gazpacho

Place 2 pounds ripe tomatoes, 1 large, seeded green pepper, 2 cloves garlic, a thick slice of bread which has been soaked in water, and 1 onion in a food processor or blender. Process until smooth. Stir in ½ cup extra-virgin olive oil and a little vinegar and seasoning to taste. Chill until required. Serve with small bowls of chopped tomato, cucumber, green pepper, onion, and hard-boiled eggs for diners to help themselves.

Product Box

Monte del Duca extra-virgin olive oil
Monte Trazzonara extra-virgin olive oil
Antica Fattoria Caroli extra-virgin olive oil
Caroli extra-virgin olive oil
Flavored extra-virgin olive oils: truffle, chili, pepper, garlic, rosemary, porcini, mushroom, oregano, basil, and sage

Casino di Caprifico

This is an organic extra-virgin olive oil from the hills of the Abruzzi on the Adriatic coast of Italy.

Giacomo Santoleri farms land around Chieti, just inland from the coastal town of Pescara. He concentrates on producing organic olive oil and organic wheat. The olive groves are situated on the hills of Guardiagrelle outside the town. Here he grows a mixture of Leccino, Gentile di Chieti, and Intosso olives. The olives are carefully selected and pressed in the traditional way. The oil is unfiltered.

The result is a very mild and sweet oil with a lightly grassy aroma. The taste is mild and unctuous with eggy flavors moving to more grassy tones and then a light peppery finish. It is an oil which can be used for all culinary applications though the producer recommends it with pasta.

Tagliatelle with Arugula and Ricotta

Mix about 1½ cups Ricotta cheese with 6 tablespoons of olive oil. Beat until smooth. Cook plenty of organic tagliatelle in boiling water until just tender. Drain and toss in 2 spoonfuls of the cooking water and add a good handful of roughly chopped arugula and a little olive oil. Serve with the Ricotta mixture as a topping.

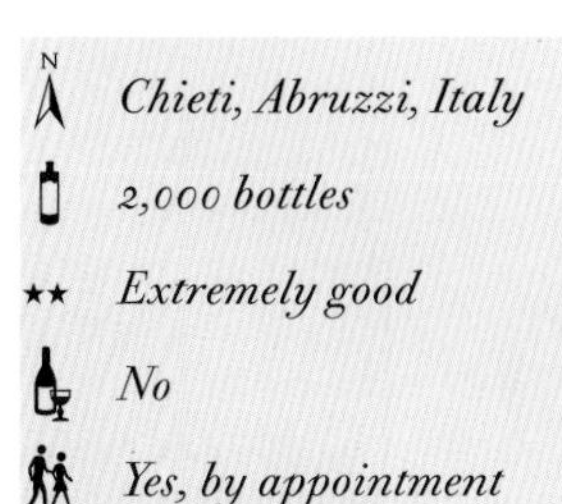

Chieti, Abruzzi, Italy

2,000 bottles

★★ *Extremely good*

No

Yes, by appointment

Castello di Ama

There are 98 acres of olive groves producing excellent extra-virgin olive oil on this estate near Siena in southern Tuscany.

The olive groves scattered among woods and vineyards in the typically Tuscan landscape of Gaiole-in-Chianti. The trees are still recovering from the frosts of 1985 but production is now on the increase.

The olives are picked by hand between early November and the middle of December. They are milled and pressed in the traditional manner and left to settle unfiltered. The oil is made up of 60 percent Correggiolo, the local type of Frantoio, and 20 percent each Moraiolo and Leccino.

The oil has a very attractive warm eggy aroma with nuts. The flavor is extremely nutty with some light chocolate tones and mild pepper. It is sweet and very lush. It is perfect as a simple dip for bread or raw vegetables. Signora Lorenza Sebasti also recommends it on Ribollita.

Above: Aerial view of this Tuscan estate.

Gaiole-in-Chianti, Siena, Tuscany, Italy

13,000–15,000 pounds

★★★ *Excellent*

Yes

By appointment only

Castello Banfi

Winner of the Leone d'Oro at Verona, this excellent extra-virgin olive oil is produced on a family-owned wine estate in Tuscany.

There are 270 acres of olive groves set in the rolling hills of the Castello Banfi estate near Montalcino in southern Tuscany. The groves are planted with Frantoio, Correggiolo, Olivastra, and Moraiolo olives.

The fruit is hand-picked in late November and early December and processed by the cold centrifugal system. The resulting oil has a complex aroma of apple skins, grass, and cut salad leaves. The flavor is lush with the bitter, sweet-sour taste of arugula and sorrel. There is plenty of pepper but the overall effect is smooth and fruity.

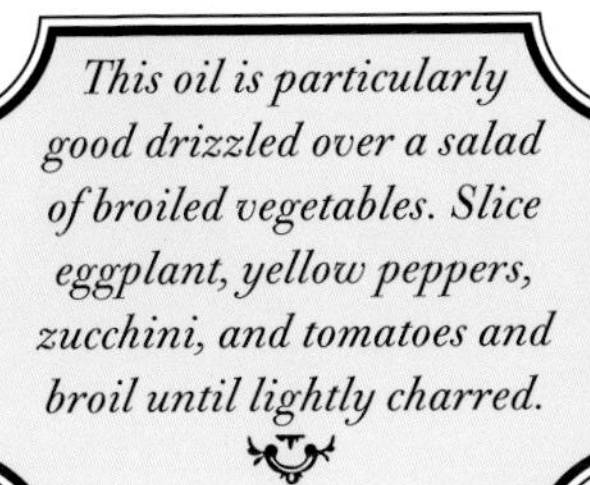

This oil is particularly good drizzled over a salad of broiled vegetables. Slice eggplant, yellow peppers, zucchini, and tomatoes and broil until lightly charred.

 *Montalcino, Tuscany, Italy*

 *5,300 U.S. gallons*

★★★ *Excellent*

 *Yes*

Yes, by appointment only

Castello Vicchiomaggio

The origins of this Tuscan estate go back a thousand years so it is no surprise to find that it produces a typically Tuscan style of extra-virgin olive oil.

The Castello Vicchiomaggio started off life as the Castello Longobardi and did not get its present name until the Renaissance when it was used for festivals to mark the month of May – *maggio* in Italian. Today the estate is owned by John Matta, but responsibility for the olive groves lies in the hands of PierPaolo Brandani, whose father held the same position before him.

The olive trees in this part of the world are quite small and one tree yields only enough olives to make one quart of oil. So production is small but of high quality. The oil smells typically grassy with apples and salad leaves but the taste takes in bitter chocolate as well as arugula,

Above: The beautiful buildings of the Vicchiomaggio estate.

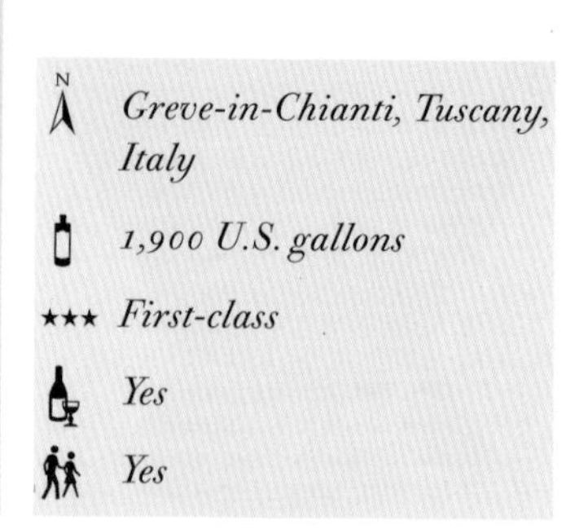

Above: Overlooking the olive groves towards the villa.

sorrel, and watercress. There is plenty of pepper too. Some people will love its distinctive character, others may prefer something a little less assertive.

If you would like to taste the oil on site there is a tasting facility for both oil and wine which is open all year round. The estate is situated in typical Tuscan countryside at Greve-in-Chianti about halfway between Florence and Siena. You can also eat at the restaurant between Easter and November or rent a holiday apartment and experience the Tuscan lifestyle firsthand.

Colonna

A great deal of care goes into all the aspects of production on The Colonna estate in southern central Italy and the result is a product of real quality.

The Colonna estate in Molise, just south of Rome, was owned by Prince Francesco Colonna and is now being run by his daughter Marina Colonna. Olives are one of the main crops here, and unlike many of the leading estates producing olive oil, grapes are nowhere to be seen.

The groves are planted in hilly country near the small town of San Martino. A number of different varieties are grown including Coratina, Leccino, Ascolano, Nocellara, and Peranzana. The estate is also experimenting with some non-Italian olives such as the Kalamata.

One of the secrets of the consistent quality of the Colonna extra-virgin olive oil is that the different varieties are harvested, milled, and pressed separately. They are then blended to achieve an oil which retains its excellence

Above: Olive groves on the Colonna estate.

Bosco Pontoni, San Martino, Molise, Italy

7,900 U.S. gallons

★★★ *Excellent*

No

Yes, by appointment

Above: The beautiful Colonna villa.

every year. Picking starts in the middle of October and goes on until December. Most, though not all, of the olives are harvested by machine just before they are fully ripe. They are always processed within 24 hours.

Once the olives have been milled to a paste they are pressed in a special press which uses neither heat nor pressure. Instead the press works on the principle of percolation, exploiting the difference in the surface tension between oil and water. The olive paste is passed over thousands of shifting stainless-steel blades in a Sinolea machine. The oil adheres to the blades and is funneled off. The result is an oil of exceptional purity.

Colonna extra-virgin olive oil has a wonderfully fresh and penetrating aroma with a strong smell of fresh olives. There are also overtones of tomato skins, lemons, and tart apples. The taste is grassy but rounded and oil has a peppery aftertaste. Colonna oil is often cited as a benchmark for southern Italian oils, but in some ways it is more Tuscan in style than southern. It certainly tastes "greener" than many other oils from the south.

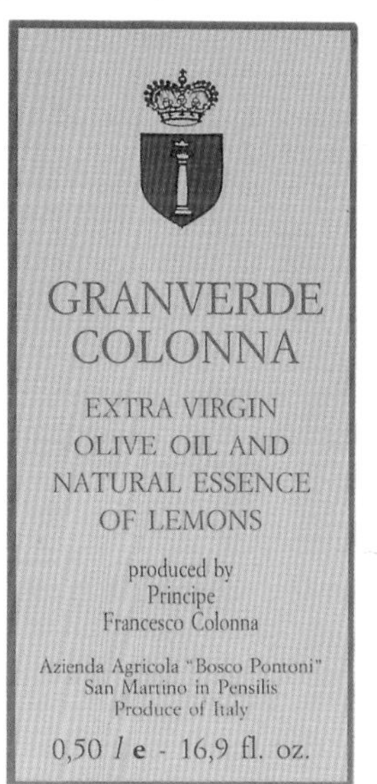

A product which used to be made exclusively for family use is now so popular that thousands of gallons are sold every year. It is Granverde, a lemon-flavored extra-virgin oil. This is made by adding a percentage of untreated fresh lemons to the olives before they are milled. The result is an oil of intense lemon flavor.

Mandarino or mandarin-flavored oil and Arancio or orange-flavored oil have recently been added to the range. They, too, have a great intensity of flavor.

The Colonna estate also produces first-class table olives. There are large green Bella Cerignola olives which are fresh-tasting and crunchy in texture. These are packed in brine. Quite different are the spicy black Coratina olives. These are brushed with oil and packed with chilies. The effect is quite hot.

Lemon Salad Dressing

Mix a little good-quality balsamic vinegar with lemon-flavored olive oil, salt, and pepper. Use to add interest to a plain green salad.

Product Box

Colonna extra-virgin olive oil
Colonna Granverde – lemon-flavored extra-virgin olive oil
Colonna Mandarino – mandarin-flavored extra-virgin olive oil
Colonna Arancio – orange-flavored extra-virgin olive oil

Large green table olives
Spicy black table olives

COLUMELA

This Spanish range of extra-virgin olive oils takes its name from the Andalusian sage and agriculturalist who lived in Rome in the first century A.D.

Lucio Julio Moderato Columela wrote a treatise on olives and agriculture covering all aspects of cultivation and production and surprisingly little has changed since those days. The principles are the same even if the equipment is more modern.

Columela oils are produced in southern Spain from Picual and Hojiblanca olives. The range includes a blend of the two as well as oils from each individual variety.

Columella extra-virgin olive oil has a very fresh and attractive aroma of lemons, melons, and tropical fruit. The flavor is fruity with spicy almonds and a very peppery punch which fades to a fruity aftertaste.

The single-variety Picual extra-virgin olive oil has a lighter aroma of melons and almonds and a very smooth and sweetly nutty taste with medium pepper. The flavor of this oil is quite complex with a touch of chocolate among the nuts and an interesting finish with sweet fruit.

The Hojiblanca single-variety oil is softer and less definite than the Picual with easy fruity flavors and medium pepper.

N *Córdoba, Andalusia, Spain*

Not known

★★★ *First-class*

No

No

Katsouris Brothers

Cypressa and Eleanthos

The Katsouris brothers turned to shipping olive oil from the Greek mainland when the family's olive groves in Cyprus were lost.

Cypressa is the brand name for extra-virgin olive oil produced from Kalamata olives grown in the Lakonia region of southern Greece. The fruit is hand-picked from October to January and processed by the cold centrifugal system. The oil has an aroma of cut leaves and nuts. This is echoed in the flavor with good pepper and a sweetly nutty aftertaste.

The second oil imported by this company from Greece is Eleanthos. This extra-virgin olive oil has an attractive aroma of cut grass and salad leaves with nutty undertones. The flavor is quite complex with cut grass and toasted nuts. There is medium pepper with an excellent hazelnut flavor on the aftertaste with just a touch of bitter-sweet sorrel.

Product Box

Extra-virgin olive oil

Cracked green olives marinated with olive oil, lemons, and herbs

London, U.K.

Not known

★★ Very good and extremely good

No

No

Greek Olive Oil and Hazelnut Cookies

Beat 1/3 cup extra-virgin olive oil with 1/4 cup sugar, juice of 1 orange, and a little brandy. Add this mixture to 1 cup all-purpose flour and 1/4 cup ground hazelnuts and 1/2 teaspoon cinnamon. Mix to a firm but soft dough, adding a little water if required. Shape into flattish cookies and bake in a moderate oven for 15–20 minutes until well-browned and done.

Fattoria
Dell'Ugo

The Amici Grossi family have farmed the Dell'Ugo estates in Tavernelle-Val-de-Pesa for six generations.

Frantoiano is the predominant olive variety on this estate which is situated about halfway between Florence and Siena in central Tuscany. It accounts for 80 percent of the blend, with Leccino and Morellino making up the rest. The olives are picked by hand in November and pressed in the traditional manner at the local village press. The oil is then stored in traditional containers under the house.

Dell'Ugo extra-virgin olive oil is apple-fresh with an aroma of cut grass and nuts. The taste is fruity with developing filberts and a rich bitterness before the peppery aftertaste. Serve quite simply, in the Tuscan manner, as a dip for crusty bread or use to dress hot vegetables such as green beans, artichoke hearts or broccoli with almonds.

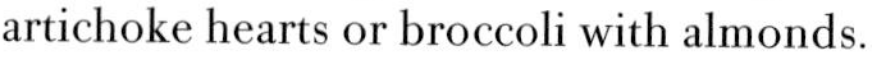

The villa on the Dell'Ugo estate.

Tavernelle-Val-De-Pesa, Tuscany, Italy

44,000 pounds

★★ *Extremely good*

Yes

Yes, by appointment

Azienda Agricola Acconia

DONNAVASCIA AND FAVARELLO

These two sweet and fruity extra-virgin olive oils are produced by the Bevilacqua family who have farmed the land in this part of Calabria since the early sixteenth century.

The farm is located on flat ground not far from the Tyrrhenian coast of southern Italy. The groves are filled with very large Carolea olive trees, characteristic of Calabria.

The Azienda Acconia boasts a veritable museum of traditional equipment including old mills, presses, and terracotta storage jars, but today the oil is produced with much more modern equipment. Because of the size of the trees the workers are able to use machines for some of the harvesting and the olives are processed by the centrifugal cold separation process. However, quality remains the goal.

Donnavascia is full of the scent of freshly pressed olives and grated apples. It has a fruity flavor with a touch of toffee bitterness and light to medium pepper.

The Favarello has a similar aroma but is much sweeter and fuller with lush fruit and just a touch of pepper. The olives for this oil are probably picked a little later in the season.

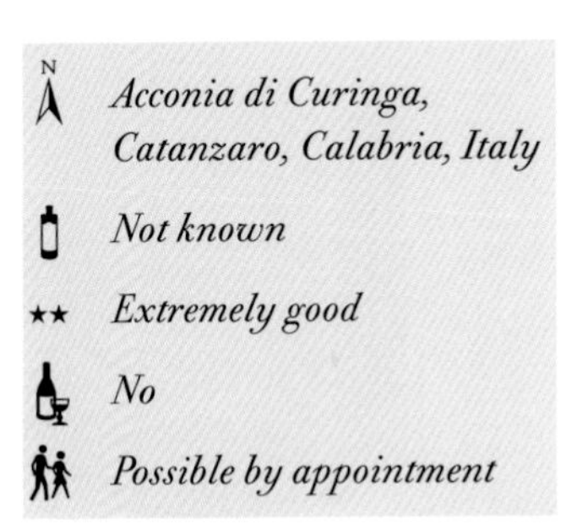

Fattoria dei Barbi

In 1985 the frost devastated the olive groves of the Fattoria dei Barbi but they were replanted and the extra-virgin olive oil now comes from very young olive trees.

The ancestors of the present owners of the estate produced wines in the sixteenth century but it has only been in the last hundred years or so that they have worked on improving the quality of the wines and concentrating on the production of these, as well as olive oil and cheese.

The new olive grove near Montalcino in southern Tuscany is made up of 90 percent Frantoio and Correggiolo. The latter variety is closely related to Frantoio and in this area it is sometimes difficult to sort out the difference. The remaining ten percent are Leccino and Impollinatori. The fruit is hand-picked in November and December and it is pressed in the traditional manner at the oil mill at nearby Petroio.

When the oil has just been pressed it has a wonderful aroma of tomato skins and warm almonds. It is thick and smooth with quite a peppery finish. It tastes of olives and nuts and this latter component intensifies a month or two after the oil has been pressed.

This oil is particularly good served with a warm salad of green beans or drizzled over broiled trout.

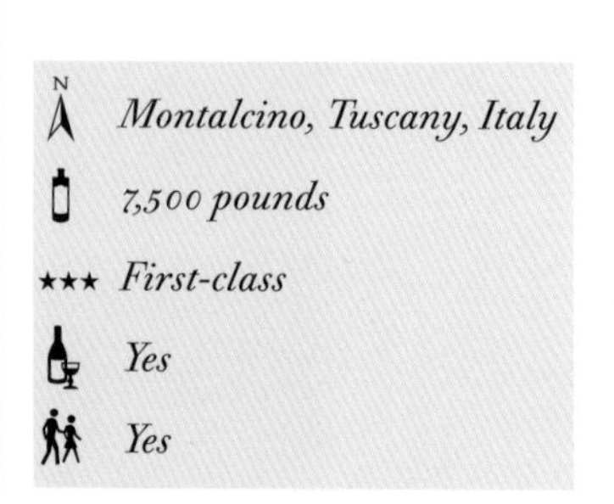

Fattoria di

Felsina

Felsina extra-virgin olive oil is extracted by one of the oldest mills still in use in Tuscany. Located by the Castello di Farnettella in Sinalunga, it still has its original stone mills and free-standing presses.

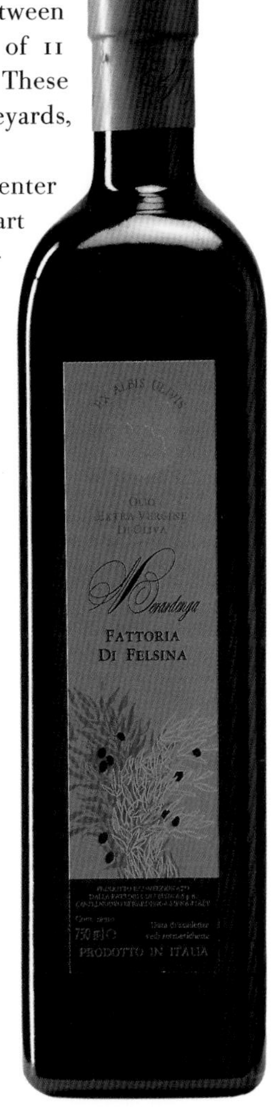

The Felsina estate is situated in the Castlenouvo Berardenga district between Siena and Lake Trasimeno. It consists of 11 farms which date back to medieval times. These farms were bought, mainly for their vineyards, in 1966 by Domenico Poggiali.

The olive groves are planted in the center of the Felsina estate and for the most part have been planted to face south-southwest. Frantoio, locally known as Correggiolo, olives account for 90 percent of the Felsina oil. The other ten percent is made up of Moraiolo, Leccino, and Pendolino.

The harvest takes place at the end of October and the early part of November. Picking is done entirely by hand and the olives are milled and pressed traditionally. The oil is not filtered.

Felsina olive oil has a sweetly fruity aroma and flavor with lots of grassy nuttiness. The pepper is light with a velvety finish.

Sinalunga, Siena, Tuscany, Italy

2,600 U.S. gallons

★★ *Very good*

Yes

Yes, by appointment only

California Olive Oil Company

Figone's

The Figone family emigrated from Italy to California's San Joaquin Valley in the 1800s but it was not until 1991 that they decided to install an olive mill on the estate.

Today Frank J. Figone presses extra-virgin olive oil from a mixture of 60 percent Mission and 40 percent Manzanillo olives. The Manzanillo olives come from a small grove on the estate.

The olives are harvested by hand from October through January. The fruit is then crushed and hydraulically pressed in the traditional manner. The result is an interesting oil with an aroma of apples and figs. It tastes of black olives with a touch of pepper.

Cheese with Olive Oil

Serve thin slices of hard cheese such as Pecorino drizzled with extra-virgin olive and freshly ground pepper or chopped herbs and plenty of country bread.

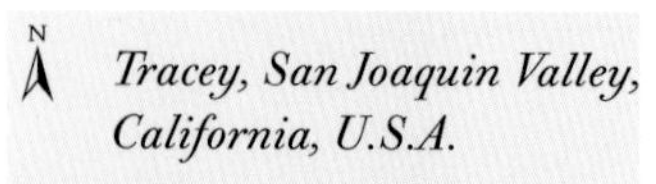

Tracey, San Joaquin Valley, California, U.S.A.

1,500–3,000 U.S. gallons

Good

No

Yes, by appointment

Filippo Berio

There really once was a Señor Filippo Berio, but the name has now become a trademark of one of Italy's leading oil producers.

Filippo Berio lived and worked in Lucca in the nineteenth century, buying olive oil from the local growers and selling it to Italian immigrants in northern Europe and in the U.S. In 1892 he joined Giovanni Silvestrini who was one of the founders of Salov, together with the grandfather of the current directors of the company. The Berio family ceased to exist at the beginning of the century.

Today the company buys olive oil from all over Italy and beyond, and blends a range of popular olive oils which sell around the world.

Filippo Berio extra-virgin olive oil has a fruity aroma with a lemon grassiness. The flavor is similarly fruity and slightly meaty with nuts and bitter leaves. There is a strong peppery punch which builds up on the finish.

In some years the company releases a "nuovo" extra-virgin oil which is quickly pressed and bottled from early-picked olives. If you like very bitter and punchy olive oil this is the one for you.

Tenuta Agricola

Fontodi

The name Fontodi dates back to the Etruscans and before. This ancient estate near Greve-in-Chianti, just south of Florence, produces good extra-virgin olive oil and Chianti Classico wines.

The Manetti family bought the estate in 1969 and set about improving both the olive groves and the vineyards. Today the estate is known for its quality produce including what is considered to be one of the best super-Tuscan wines, Flaccianello.

The olive groves are planted with almost 100 percent Frantoio olives which are hand-picked in November. The fruit is processed by the cold centrifugal system.

The oil has a typically strong grassy, aroma of salad leaves with a similar flavor which includes bitter arugula and aggressive pepper which settles into a smooth and leafy aftertaste. The best way to use this oil is in the local Pinzimonio (see page 58). It is also very good stirred into the traditional robust Tuscan vegetable or bean soups.

N *Greve-in-Chianti, Panzano, Tuscany, Italy*

800 U.S. gallons

★★ *Extremely good*

Yes

Yes, by appointment

Tenuta di FORCI

This ancient estate set in the hills northwest of Lucca was, like so many other estates in the area, badly hit by the frosts of 1985.

However, production of extra-virgin olive oil is coming back on stream now thanks to the efforts of Baronessa Luisa Diamantina Scola-Camerini who now owns the estate. She is working on restoring the ancient olive groves to their former glory.

Frantoio is dominant in these groves at 75 percent. Leccino makes up a further 15 percent. The rest of the trees are Maurino, Moraiolo, Pendolino, and a mixture of very local varieties. The harvest takes place in November and December and both traditional and centrifugal processes are used.

This oil has a nutty aroma with undertones of coffee and chocolate. The flavor is full and fruity with medium pepper.

PRODUCT BOX
Extra-virgin olive oil

Marinated olives
Olive paste

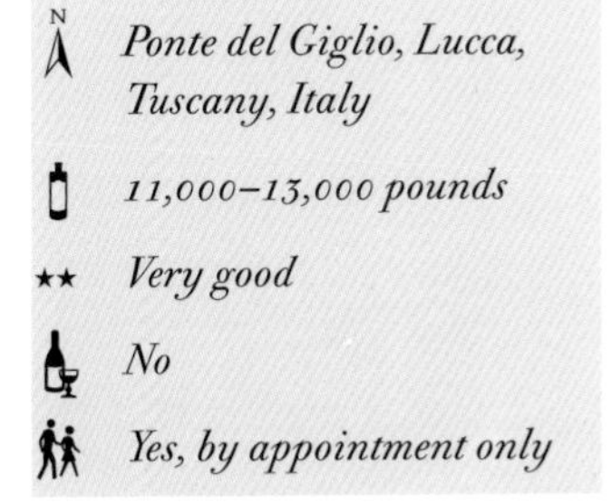

Frantoio di Santa Tea

Fruttato-Intenso and Dolce-Delicato

History and technology go hand-in-hand at Santa Tea and the result is superlative extra-virgin olive oil.

The land was originally owned by the Carmine Monastery in Florence and there are records of its sale to the Gonnelli family in 1585. The farm, now owned by the Bettali family, is situated in the parish of Santa Tea near Reggello on the eastern hills of the Arno Valley not far from Florence.

Years of experience in the cultivation of olives are now complemented by modern scientific methods. Carolea olives are the chosen variety. The trees are planted at about 1,200 to 1,300 feet in an area which is sheltered by the Vallombrosa hills.

The olives are harvested at different stages in their development to produce oils with different characteristics. In November the first batch of olives is harvested when they are not fully ripened and are therefore still fairly green in color. The oil from these olives is

- *Santa Tea, Reggello, Tuscany, Italy*
- *8,200 U.S. gallons*
- ★★★ *First-class*
- *No*
- *Yes, by appointment only*

labeled Fruttato-Intenso and you easily pick the right bottle as the illustration on the label shows green olives.

This oil is very intense with a strong aroma of fresh olives, salad leaves, and cut grass. The taste is equally strong with flavors of arugula and sorrel. This bitter, almost chocolaty flavor moves on to a peppery kick followed by a really sweet and smooth aftertaste. It is an oil for strongly defined tastes such as game, roasts, and winter vegetables. It also goes well with arugula and artichokes.

Some of the olives are left to ripen on the trees until January when they are fully ripe. These olives are crushed to make the Dolce-Delicato oil. This oil is sweeter and more softly vegetal with apples and almonds but it is still quite peppery.

Santa Tea also produce another olive oil from the Frantoio olive. This has a good grassy aroma tempered by lemons and a similar taste softened by toasted nuts. It is very peppery but the aftertaste is not too fiery.

Product Box

Fruttato-Intenso extra-virgin olive oil
Dolce-Delicato extra-virgin olive oil
Cultivar Frantoio extra-virgin olive oil
Rocca di Cerbaia extra-virgin olive oil

Leccino olives in brine (green and black)

Green and black olive paste

Vegetable sauces with olive oil

Frantoio Gaziello

This small Italian family-owned business based on the Ligurian coast produces quality extra-virgin olive oils from the local Taggiasca olive.

The business started when Giovanni Gaziello introduced a water-driven press to his home village of Trucco outside Ventimiglia. Twenty years later he moved into new premises in Ventimiglia and was eventually joined by his son Giorgio.

Today the business boasts a modern stainless steel centrifugal plant, capable of processing 3,300–4,400 pounds of olives an hour. Two oils are produced and they are both made from the local Ligurian Taggiasca olive.

Gaziello Mosto is a natural, unfiltered oil made from the first olives to be picked in December and January. It has a good aroma of freshly grated apple skins, olives, and salad leaves. The flavor is much more nutty but still fresh and sweet. It can sometimes take on an almost chocolaty flavor. The texture is quite thick and lush and the wicked little peppery kick comes as quite a surprise.

Ventimiglia, Liguria, Italy

21,000–26,500 U.S. gallons

★★ *Extremely good*

No

By appointment

This oil may be packed in a large round one-quart bottle or in an attractive flask-shaped bottle with a cork stopper which is attractive to display in the kitchen.

Frantoio Gaziello is the company's second oil. It, too, is unfiltered. It is made from olives from the February to March harvest and is sweeter and a little less fruity than the Mosto. The aroma is lightly grassy. It has a sweet eggy, almost avocado flavor, which is not at all aggressive and there is very little pepper.

PRODUCT BOX

Gaziello Mosto extra-virgin olive oil

Frantoio Gaziella extra-virgin olive oil

Pickled black table olives

Pickled and pitted and packed in olive oil

Black olive paste or tapénade

The Frantoio Olive Oil Company

Castellare di Ugnana

This enterprizing company has a foot in both Italy and California, offering extra-virgin olive oil from their own estate in Tuscany and a Californian oil which is pressed on display at the Frantoio restaurant in Mill Valley.

Having successfully developed their own Tuscan olive oil, the Zecca family decided to take their expertise to California.

Rather than start up a conventional plant the Zecca's decided to open a restaurant which would display the oil production during the harvest season. Mission, Manzanillo, and Ascolana olives are brought in from northern California near the Oregon border and crushed and pressed at the restaurant.

Frantoio extra-virgin olive oil is a blend of 40 percent Manzanillo fall oil and 60 percent Mission winter oil. The company also presses oils from 100 percent Ascolana, Sevillano, Picholine or Barouni olives for local estates.

The extra-virgin oil from the Castellare di Ugnana in Italy has a very light slightly earthy aroma but a distinctively attractive flavor of nutty coffee and chocolate with lemons. There is light to medium pepper and a smoothly nutty aftertaste.

Mill Valley, California, U.S.A.

6,500 U.S. gallons

★★★ *First-class (Californian oil not tasted)*

No

Yes

Frescobaldi

Laudemio and Castelgiocondo

The Marchese de' Frescobaldi, founder and president of the Laudemio marketing concept, produces these extra-virgin olive oils from olives grown on the Frescobaldi estates in Tuscany.

The aristocratic Frescobaldi family can trace its history back to the twelfth century and before. They have been politicians, artists, explorers, and even bankers to the English throne. During the thirteenth century they lent large sums of money to Edward II and his father and Berto Frescobaldi became a Crown Counsellor.

A hundred years later the family were asked to leave and their wealth was confiscated. In retaliation the Frescobaldi stopped the shipments of their own wine and bought up all stocks of French wine in Bordeaux. The English court was without wine for some time!

Product Box

Frescobaldi Laudemio extra-virgin olive oil
Castelgiocondo extra-virgin olive oil

Florence, Tuscany, Italy

4,600 U.S. gallons Laudemio and 920 U.S. gallons of Castelgiocondo

★★★ *First-class*

Yes

Yes, by appointment only

Above: The approach road to the Castello Nipuzzano estate.

For centuries the family estates have produced top-quality wines which were famous well beyond their own locale but it has only been in more recent years that their olive oils have been available to a larger audience. One oil is now marketed under the Laudemio brand name; the other is sold under the name of the estate from which it comes.

Frescobaldi Laudemio is made mainly from olives from the Castello Nipozzano estate in the Chianti Rufina area of Tuscany near Pelago. However, some olives do come from the other family estates of Camperti and Poggio a Remole in the Chianti Rufina area, Castello di Pomino in Pomino, and Castiglioni and Montagnana in the Chianti Colli Fiorentini area. They are all within easy reach of Florence.

The blend of olives is 80 percent Frantoio, ten percent Moraiolo, and ten percent Leccino. The groves were badly damaged by the frosts of 1985 and some of them have been replaced with modern drip-irrigated, high-density systems. The remaining groves have been rejuvenated from old trees and interplanted with young trees to increase plant density and reduce labor costs.

The groves enjoy a very dry but cool location which is almost constantly swept by northeasterly winds. This keeps the olive fly away and insures that the olives are not very ripe when they are harvested in the early days of November. All picking is completed by the end of the month. The oil is produced with the modern continuous cycle method and is stored in terracotta pots where it naturally decants.

The result is a wonderfully vibrant green color and an elegantly aggressive aroma and flavor. It smells and tastes of freshly-pressed olive fruit, cut grass, and crushed sorrel leaves. It is very definitely peppery with a bitter sweet aftertaste.

It dominates delicately-flavored food but gives immediate life to more boring ingredients. It makes a great marriage with other distinctive flavors such as chicken livers, asparagus with Parmesan, or barbecued steaks.

Castelgiocondo is pressed in the same way as Laudemio but is made from a different blend of olives grown on the Castelgiocondo estate in Montalcino in southern Tuscany. The groves are mostly new and have been planted with the new high density systems used for Laudemio. The countryside here is much gentler and the environment considerably warmer than in the Florence district. By November the olives are slightly riper than those growing in groves further north, though still green.

The result is an oil which is more mature though only a little less peppery. The aroma is strong with fresh olives, apples, and cut salad leaves. The flavor is less intense but still fruity with plenty of pepper in the aftertaste.

Above: The Frescobaldi family.

Fresh Olive Company of Provence

This company's French blended extra-virgin olive oil is made from ripe Provence olives picked mainly in December.

The blend is based on oils produced by various small farmers and cooperatives in the Gard, Bouches du Rhône, and Hérault regions of central southern France. Much of the pressing is done in the traditional manner by a small family business in the Gard.

The olives include Picholine, Beruguettes, Grossane, Salonenque, and Beldi and the percentages vary from year to year, depending on the harvest.

The oil has a lightly grassy aroma with lemony overtones and tastes very sweet and fruity. It has a surprisingly peppery kick to it though the aftertaste is quite smooth. It is a mild and versatile oil.

Product Box

Extra-virgin olive oil

Aromatized olives, wide range

Black olive paste

N *Acton, West London, U.K.*

Not known

★★ *Very good*

No

No

Fattoria di
Fubbiano

This award-winning extra-virgin olive oil comes from olives which are fertilized biologically by 20 Haflingen horses which are allowed to graze the groves.

The estate is situated in the hilly countryside between the village of Tofori and San Gennaro near Lucca in Tuscany. At the heart of the estate is an attractive villa surrounded by the old wine cellars, oil press and bakehouse. Visitors can stay in the main villa or in one of the small farmhouses set amid the vineyards and olive groves.

The groves are mainly planted with Frantoio though there are a few Moraiolo, Leccino, and Pendolino trees mixed in. The trees are trained into the typical vase shape of the region. The fruit is hand-picked and pressed traditionally. The oil has a very attractive aroma of freshly grated apple skins. Its taste is similar, with an interesting touch of bitterness leading to light to medium pepper. The oil is very sweet and smooth.

The oil makes an excellent salad dressing mixed with a little white or red wine vinegar. Do not use too much vinegar or the apple-fresh flavor of the oil will be overwhelmed. It also makes a good mayonnaise.

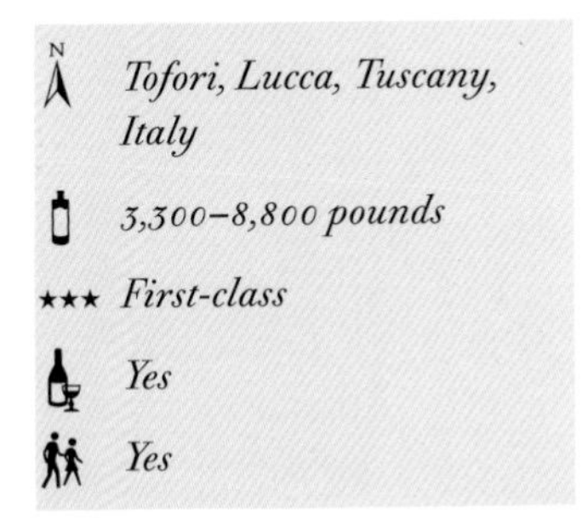

Olivar de Segura

Fuentebuena and Oro de Genave

These two extra-virgin olive oils carry the Denomination of Origin seals from the Sierra de Segura region of southern Spain.

The dominant olive variety in the rugged hills of the Sierra de Segura region is the Picual and to qualify for the D.O.C. label all the olives must come from farms within the region.

The Oro de Genave only uses olives grown in groves run on organic principles. The fruit is harvested by hand and both traditional and centrifugal methods are used to extract the oil. The extra-virgin olive oil has a rich aroma of lemons, apples, melons, and nuts. The flavor is full of warm melons and lemons with a touch of cut grass. The lightish pepper builds up but finishes very sweetly.

Fuentebuena extra-virgin olive oil is much more nutty in character with lemons and almonds dominating the aroma. The flavor, too, is full of aromatic nutty tastes with bittersweet leaves moving to medium pepper and a smooth and sweetly nutty aftertaste.

N *Puente de Genave, Jaén, Spain*

not known

★★★ *First-class*

No

No

Above: A scene typical of the rugged countryside of this region of southern Spain.

Gazpacho

Place 2 pounds ripe tomatoes, 1 large, seeded green pepper, 2 cloves garlic, a thick slice of bread which has been soaked in water, and 1 onion in a food processor or blender. Process until smooth. Stir in ½ cup extra-virgin olive oil and a little vinegar and seasoning to taste. Chill until required. Serve with small bowls of chopped tomato, cucumber, green pepper, onion, and hard-boiled eggs for diners to help themselves.

Oleificio

GABRO

This well-flavored extra-virgin olive oil from Calabria is a winner both with its devotees and in competition.

Until quite recently much of the Gabro oils and table olives were sold to large bottling companies producing national brands of olive oil. However, Mario Brogna, the current owner of the company, saw a growing market for superior-quality olive products and decided to develop his own brands. He is now one of the largest private organic oil producers in Italy. His oil is accredited by the A.I.A.B., the Italian Association of Organic Agriculture.

The olives are all grown on the family estate in the Calabrian hills in an area known as Calabria Citra. The mix of olive varieties is 70 percent Tondino and 30 percent Grossa di Cassano. The fruit is picked by hand. Unripe olives are pressed in the traditional manner; ripe olives go through the centrifugal presses.

N *Cosenza, Cassano Ionio, Calabria, Italy*

5–6 tons

★★★ *Excellent*

No

Yes

The oil has a delicate but definite aroma of freshly crushed olives, grass, and fruits like apples and even raspberries. The flavor is equally complex. It is full of fruit and nuts and, of course, olives with light to medium pepper. The aftertaste is sweet.

Mario Brogna also makes intense herb-flavored oils from Carolea and Nocellara olives. The fresh herbs are crushed with the olives while the olive oil is being made.

Spaghetti alla Norcina

Cook about 1 pound spaghetti in boiling salted water until just tender or al dente. *Meanwhile bone, wash, and chop two salted anchovies (or four to five canned anchovies in oil) and place in a pan with six tablespoons of olive oil and 1 minced clove garlic. Sauté gently and add as much black truffle as you can afford. Continue cooking for two minutes. Drain the pasta thoroughly and toss in olive oil. Pour the truffle sauce over the top.*

Product Box

Oleificio Gabro Organic extra-virgin olive oil

Fruttato extra-virgin olive oil

Sybaris extra-virgin olive oil

Flavored herb oils

Olive pâté made from Cassano and Corniola olives

Gaea

Trading

This company, named for the mythological Greek mother earth, selects its almond flavored extra-virgin olive oil from six villages in the Greek province of Etoloakarnania.

The villages are Makinia, St. Thomas, St. George, Eratini in the Makinia region which is situated on the Greek mainland north of the Gulf of Patras and Skala and Yomvoku in the Lepanto region a little further north.

The timing of the harvest is as precise as the attention to detail and usually takes place between 20 November and 10 January. There are both traditional and centrifugal presses in the region but Gaea prefer those which have been pressed traditionally.

The local olive varieties are Ladoelia and Koronia and they are usually pressed in equal quantities. The resulting extra-virgin olive oil has a lightly grassy aroma and a nutty, fruity flavor with light pepper.

Product Box

Extra-virgin olive oil

Marinated olives in olive oil
Kalamata olives in paprika
Mixed olives with lemon and oregano

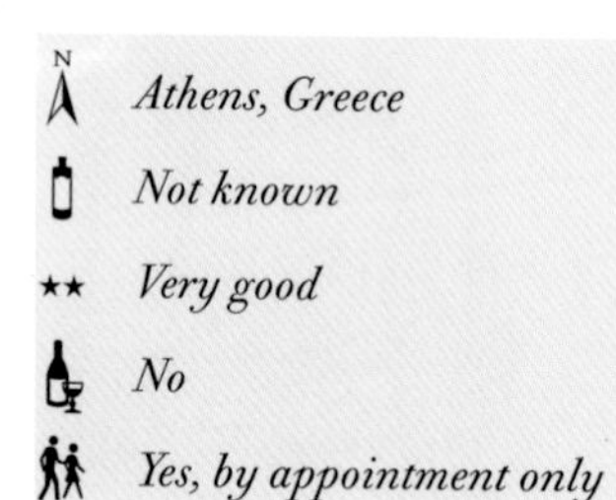

Athens, Greece

Not known

★★ *Very good*

No

Yes, by appointment only

Niccolini di Carlo Niccolini

Gemma

This rustic extra-virgin olive oil, which tastes of ripe black olives, comes from groves near Calcione in southern Tuscany.

The Niccolini business is based in Calcione only a stone's throw from the Florence–Rome highway. They are merchants rather than growers and source their oils from the estates in the surrounding hills. The company buys first-class oils and packs them for the world market. They are members of the prestigious Mastri Oleari.

Gemma is pressed from a mix of Leccino and Frantoio. It has the aroma of fresh olives, apples, and grass typical of many Tuscan oils. Its flavor is more unusual, offering a strong taste of black olive, salad leaves, and artichokes. There is plenty of peppery bitterness in this very assertive oil.

Product Box

Gemma extra-virgin olive oil
Frantoio extra-virgin olive oil
La Nonna extra-virgin olive oil

Calcione, Arezzo, Tuscany, Italy

10,500 U.S. gallons (Gemma alone)

★★ *Very good*

No

Yes

Grappolini

Three generations of experience go into the selection of the Grappolini range of extra-virgin olive oils.

Each Grappolini oil is typical of its region and suited to a different culinary use. So far I have only sampled two of these oils. Il Gentile has a flavor similar to sweetly aromatic salad leaves and a good peppery punch leading to a smooth and fruity aftertaste. The flavor of Suavis is an unusual herbaceous mix of salad leaves and wet hay with nutty almonds and good pepper.

Product Box

Ampolla extra-virgin olive oil
L'Orchetto extra-virgin olive oil
Il Gentile extra-virgin olive oil
Paneolio extra-virgin olive oil
I Toscano extra-virgin olive oil
Il Pregiato extra-virgin olive oil
Il Giovane extra-virgin olive oil
Suavis extra-virgin olive oil

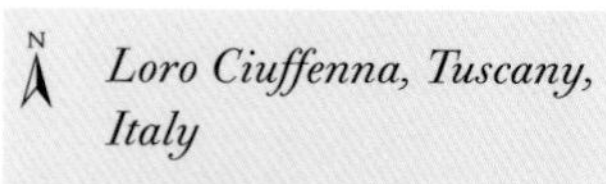

Loro Ciuffenna, Tuscany, Italy

Not known

★★ *Extremely good*

No

No

Harrison Vineyards

A one hundred-year-old Italian granite olive crusher processes the olives from groves near to this small wine estate in Pritchard Hill, California.

Lyndsey and Michael Harrison thought about living in Tuscany before they bought this beautiful piece of land in the Napa Valley in 1988. They had no intentions of farming but they now make both extra-virgin olive oil and Cabernet Sauvignon and Chardonnay wines.

The wines came first as, with the help of local neighbors, they decided to put existing vineyards back to work. On their frequent trips to Italy the Harrisons noticed that many of the wine estates had olive trees and vines growing side by side and they resolved to add olive oil to their repertoire.

They bought an old granite olive crusher in Italy and shipped it back to the U.S. where they now crush olives from various nearby groves. The oil is made up of approximately 50 percent Mission olives, 30 percent Manzanillo and 20 percent Picholine, Frantoio, Sevillano, and other less well-known varieties.

Pritchard Hill, Napa Valley, California, U.S.A.

400–600 U.S. gallons

★ *Fairly good when fresh*

Yes

No

The Harrisons' daughter Jill heads up the olive oil operation and her husband Chris Willis makes the oil. The olives are harvested by hand in late October through the middle of January, starting with very unripe olives and ending with black. They are crushed and pressed traditionally, and the resulting oil is filtered through sterile cotton.

Harrison olive oil has a warm but fresh eggy-lemon aroma. The flavor is reminiscent of black olives with a little bitter grass and peppery finish which builds up to something of a crescendo.

Warm Chanterelle and Pancetta Salad

Cook ¼lb diced pancetta in a skillet over a low heat until crisp. Remove from the pan and keep on one side. Drain away most of the fat and sauté 2 peeled and minced shallots until brown. Add ½lb trimmed and quartered chanterelles and ¼ cup pine nuts. Sauté until lightly browned and mix with the pancetta. In the same pan heat 7 tablespoons of extra-virgin olive oil and whisk in 1 tablespoon of lemon juice and 2 tablespoons red-wine vinegar. Toss plenty of mixed leaves with the pancetta and chanterelles, and the warm dressing. Serve at once with freshly ground black pepper.

Product Box

Extra-virgin olive oil
Flavored olive oil – basil and pepper

Gyfteas Group Company

Iliada

It is the Koroneiki olive which gives this extra-virgin olive oil from the Kalamata region of Greece its distinctively aromatic flavor.

Iliada is one of the leading brands of the Gyfteas Group Company which is based in the province of Messinia. This company is still family-owned even though it trades approximately ten percent of all the oil produced in this part of Greece. Its strengths lie in its ability to pick the best producers from which to buy oil, its experience in blending these oils, and its very modern quality control procedures.

Iliada is essentially a regional rather than an estate oil. The olives are grown and the oils pressed in groves and mills scattered around the city of Kalamata on the western side of mound Tayghetus in the southern central Peloponnese. Hand-picking and traditional mills and hydraulic presses are the norm here.

The oil has a particularly strong aroma of hay and dried grasses with an aromatic but fruity taste. There is more dried grass here, mixed with bell peppers and spice. It is an unusual oil but it is still very good in salads and general cooking.

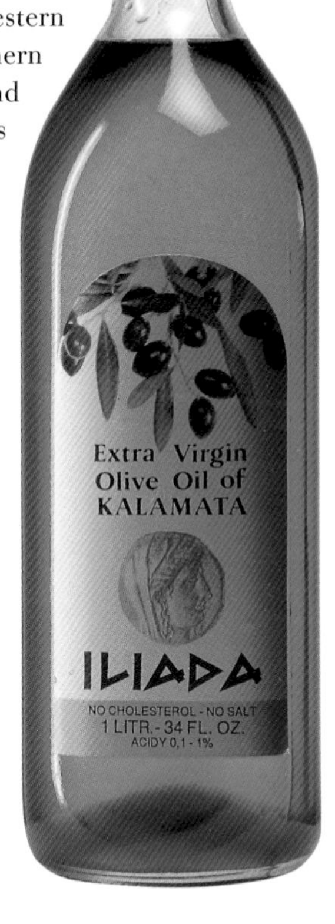

- N: *Spernogia, Kalamata, Messinia, Greece*
- *Not known*
- ★★ *Very good*
- *No*
- *Yes, by appointment*

Fattoria di Galiga e Vetrice

Il Lastro

This estate in the Chianti Rufina hills in northern Tuscany produces a distinctively peppery but elegant extra-virgin olive oil.

The olive groves of the Fattoria di Vetrice are surrounded by vineyards bursting with Sangiovese grape vines destined for the company's own Chianti wines. Almost half the estate is devoted to olives.

The predominant variety is Frantoio which accounts for 90 percent of the mix in the oil. The rest is made up of five percent Pendolino and five percent Morinello. The fruit is picked by hand and pressed in the modern plant within two days.

The oil has a very green and grassy character with an aroma of mown grass, apples, and olives. The flavor is also sharply grassy with arugula salad leaves and a peppery kick which softens a little over time. The aftertaste is smooth and elegant.

Serve this deliciously assertive oil with robust salads of warm bacon and porcini mushrooms or with chicken livers and arugula. It also makes a very good sauce for pasta if it is warmed with plenty of freshly minced garlic and a little chopped parsley or basil.

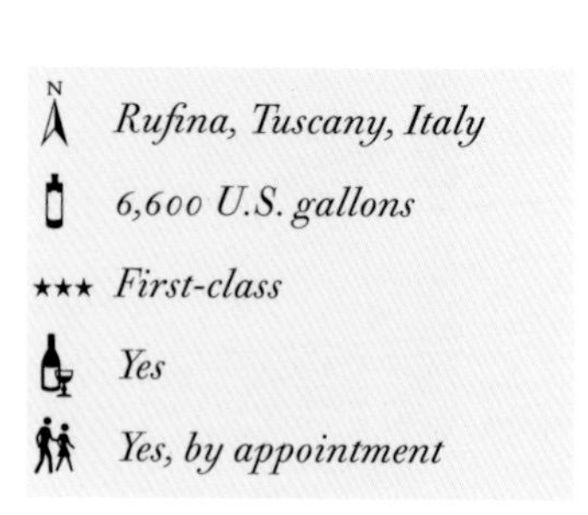

Azienda Agrobioligica

Il Poderetto

Flos Viridis and Fior del Colle

This classical Umbrian farm now owned by Dr Flavio Zaramella, President of the Corporazione dei Mastri Oleari, produces very high-quality prize-winning organic extra-virgin olive oils from a mix of traditional olives.

The foundations of the farm are built on the base of an ancient Roman tower near the medieval village of Mora near Assisi. The olive groves are set in gentle countryside of fields, woods, and streams.

The olives are cultivated organically and harvested by hand in November and December. Both traditional and centrifugal methods are used to produce the oil which is left unfiltered. It is allowed to settle for approximately three months before bottling.

Above: Dr. Zaramella's classical Umbrian farm.

Mora di Assisi, Assisi, Umbria, Italy

Not known

★★★ *Excellent*

No

Yes, by appointment

Above: The St. Francis of Assisi Basilica.

The oils are certified and guaranteed by A.I.A.B., the Italian Association for Biological Agriculture.

Four different varieties of olives are used to make Flos Viridis. The mix is 50 percent Moraiolo, 20 percent each Leccino and Frantoio, and ten percent Raggiolo. The olives for this oil are picked early in the season and the result is a grassy, fresh, and fruity oil with a touch of lemon. The olive flavor is elegant but assertive with a nicely peppery aftertaste.

Fior del Colle uses much the same mix of olives as the Flor Viridis but the Raggiolo olives are replaced by a greater percentage of Moraiolo. They are picked later in the season and the oil is even smoother and more unctuous than Flos Viridis with a very pleasant eggy and almond edge to the fresh olive fruit. It is quite peppery but finishes very sweetly.

Product Box

Flos Viridis organic extra-virgin olive oil
Fior del Colle organic extra-virgin olive oil
Il Poderetto extra-virgin olive oil
Flavored olive oils

Black olive paste
Green olive paste

Fattoria di
Isole e Olena

Set amid the rolling hills of central Tuscany this estate produces an unusual, but deliciously nutty extra-virgin olive oil.

The Isole e Olena estate is owned by the de Marchi family who bought two adjoining estates in the commune of Barberino Val d'Elsa in the late 1950s and now run them as one estate.

Around 20 percent of the estate is devoted to olive groves with vineyards and extensive woodlands surrounding them. The groves are planted with Frantoio, Moraiolo, Pendolino, Leccino, and Morchiaio but Frantoio is the dominant variety. The groves are cultivated without the use of pesticides and there is no spraying at all.

The fruit is picked by hand from early November through to the middle of December. It is milled and pressed in the traditional manner. The resulting oil has an almost meaty aroma with plenty of toasted nuts and chocolate. The flavor is warm and fruity with a velvety smoothness and medium pepper. The aftertaste is sweet and fruity.

The de Marchi family reserve this oil to serve cold with crusty bread and in salads. They will also use it in vegetable soups such as Ribollita.

Primo Estate

Joseph Foothills

Some of the very first olive groves to be planted in Australia in 1864 are the source of this extra-virgin oil from "down-under."

Along the base of the Mt Lofty range, near Adelaide, are the remains of quite large olive groves planted by the early settlers. For many years the olives were just left to fall to the ground unused.

Six years ago Joe and Dina Grilli, who were already very busy running the successful Primo Estate winery, decided to pick the olives from these and remnants of other similar groves set round the local market gardens. The result was Joseph extra-virgin olive oil.

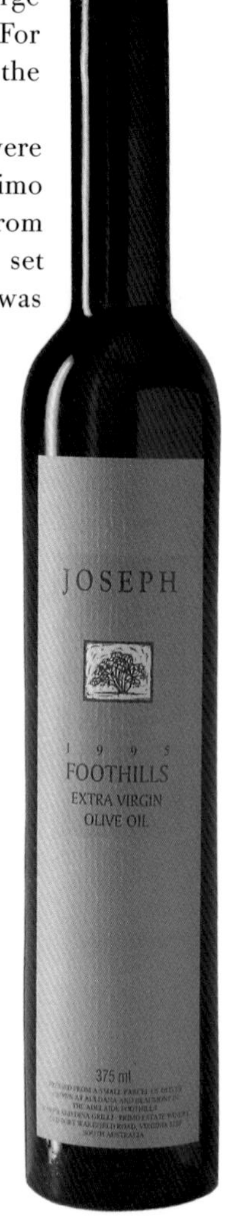

The olives, mostly Verdial, are picked before they are fully ripe when they are half green and half purple. They are crushed and pressed both traditionally and centrifugally at a nearby pressing facility. The oil is then returned to the winery where it is allowed to settle for eight weeks before bottling.

Joseph Foothills extra-virgin olive oil has a good straightforward aroma of olive fruit with lemons and grated apples. The flavor, too, is apple-fresh and fruity with a light grassiness and some warm nuts. The pepper creeps up on you unawares but the aftertaste is sweet and nutty.

Primo Estate, Virginia, South Australia

1,100 U.S. gallons

★★ *Extremely good*

Yes

Yes – June to December

Kydonia

Extra-virgin olive oil has been produced by farmers in Crete for more than 2,000 years.

This one comes from a group of farmers whose groves are situated in the hills in the northwestern part of the island. The groves are planted only with the Koroneiki olive.

The fruit for the Kydonia "special reserve" is hand-picked before it is fully ripe. The olives are crushed and pressed in the traditional manner.

Kydonia Special Reserve extra-virgin olive oil has a delicate nutty aroma and a distinctive grassy flavor reminiscent of freshly mown hay. The oil is interestingly different but quite smooth and elegant.

Taramasalata

Soak a very thick slice of bread with no crusts in water for 15 minutes. Drain and squeeze dry. Place in a food processor with ½ cup fresh or smoked, skinned cod's roe and two or three very finely chopped shallots and blend. Stir in the juice of 1 lemon and seasoning to taste. With the motor running, gradually add 5–7 tablespoons of extra-virgin olive oil.

 *Crete, Greece*

 *Not known*

★★ *Extremely good*

 *No*

 *No*

Olive Oil Cake

Blend 1 cup extra-virgin olive oil with 1 cup sugar and 2 eggs for 5 minutes in a blender. Pour into a large bowl and gradually add 4 cups flour, 2 cups raisins, $1\frac{1}{2}$ cups milk, 2 teaspoons baking soda, and 2 teaspoons ground cinnamon. Bake for $1–1\frac{1}{4}$ hours in a medium oven. Test the center of the cake with a skewer to check for doneness.

Product Box

Kydonia Special Reserve extra-virgin olive oil
Blue Label extra-virgin olive oil

Black olives with fennel
Greek olives with lemon
Kalamata olives in dill

Lake & Co.

This is a new company set up in 1996, specializing in selling extra-virgin olive oils from small producers in various parts of the Mediterranean. So far there is only one oil in the range.

This is Les Alpiles from Provence. It is a blend of local Verdale, Salonenque, Picholine, and Grossane olives which are milled and processed by the cold centrifugal method. The oil has a really fresh lemony aroma which is smooth and sweet. The flavor is lush and lemony with an interesting touch of bitterness building to medium pepper. This is a versatile oil to use with traditional Provençal dishes such as Pan Bagnat, Ratatouille, and Salade Niçoise.

Product Box

Les Alpiles extra-virgin olive oil

Tuscan extra-virgin olive oil

Black and green table olives

Black and green olive pastes

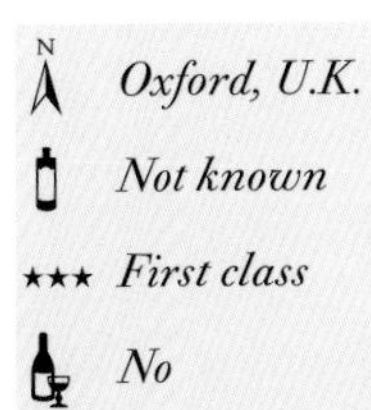

Oxford, U.K.

Not known

★★★ *First class*

No

No

Quinta de la Rosa

La Rosa

A Swedish wine estate in the Upper Douro valley near Pinhao is the source of this extra-virgin olive oil from Portugal.

The Bergqvist family have lived and worked in the center of Portugal's famous port-producing region since 1805. Until 1988 they sold their wine to other more famous port shippers but now they market their own produce direct to the customer and, of course, this includes their own estate-produced extra-virgin olive oil.

Visitors to the Douro can stay at the rambling quinta house with its bathroom carved from the rock which supports the vines above or in one of the two farmhouses on the property which have been restored.

The extra-virgin olive oil from this estate has always sold well to particular chefs who really like its earthy tastes. It is a distinctly different oil with an aromatic flavor of dried rosemary, thyme, and lavender.

N *Pinhao, Alto Douro, Portugal*

Not known

★★ *Extremely good*

Yes

Yes

Le Cesine

This distinctive southern Italian extra-virgin olive oil has only been on the market for about five years and already it is a gold medal winner.

The de Japinis family, father and son, grow their olives in the hilly countryside of Campania about 40 miles east of Naples. They are based near the small village of Monterocchetta southeast of Benevento.

A variety of different olives go into the mix including around 40 percent Ortice, 25 percent Ortolana, and 15 percent Racioppella. The remainder is taken up mainly with Leccino and Frantoio. The olives are harvested by hand in late October or early November and are processed in the traditional manner.

The oil has a wonderful aroma of olives, apples, and tomatoes. These flavors are intensified in the mouth, leaving an interesting touch of sorrel leaves on the palate. There is very little pepperiness.

Serve on simple salads and fresh vegetables to show off the complex flavors of the oil or warm very gently and drizzle over poached fish with capers or lightly broiled chicken breasts with sun-dried tomatoes.

Monterocchetta, S. Nicola Manfredi, Benevento, Campania, Italy

1,900 U.S. gallons

★★★ *Excellent*

No

By appointment

Frantoio Eugenio Marco

LE FASCE AND CEPPO ANTICO

This small Ligurian family business produces consistently good extra-virgin olive oils which are quite long-lasting.

Local Taggiasca olives are the basis of this family's range of oils. They are gathered from terraced olive groves which ascend almost 2,000 feet up the steep-sided mountain ravines of western Liguria. You can see the pickers precariously negotiating the heights in January and February.

Frantoio Marco produce three oils from these olives and they are typical of the region. They have a warm and grassy aroma with overtones of apples and almonds. The Le Fasce oil, which is unfiltered, has a rich eggy aroma and taste. The oils are luscious in texture and taste sweetly nutty with light to medium pepper.

Liguria is the home of Pesto Sauce for pasta. It is made with olive oil, basil, pine nuts, and cheese and these oils are the obvious choice. They also taste very good with the region's abundant seafood.

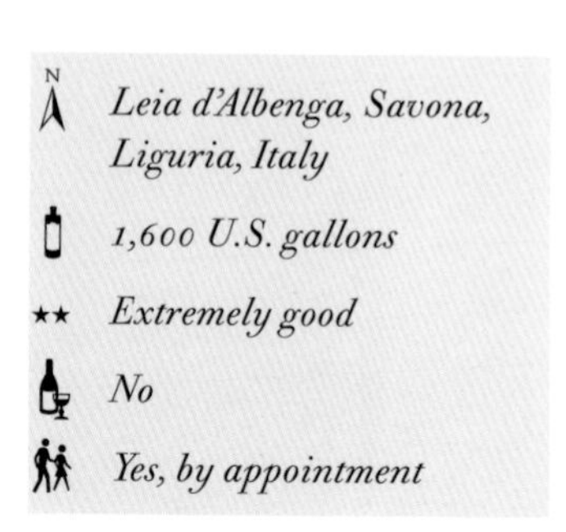

Leia d'Albenga, Savona, Liguria, Italy

1,600 U.S. gallons

Extremely good

No

Yes, by appointment

Fish Baked in Foil

Take any small white fish which has been gutted and fill the cavity with a bunch of rosemary, 1 sliced clove garlic and some salt. Add 1 table spoon of olive oil and wrap in a foil parcel. Bake at 325° F for about 25 minutes. When it is done, skin and bone it, add some fresh olive oil as a sauce, and serve at once.

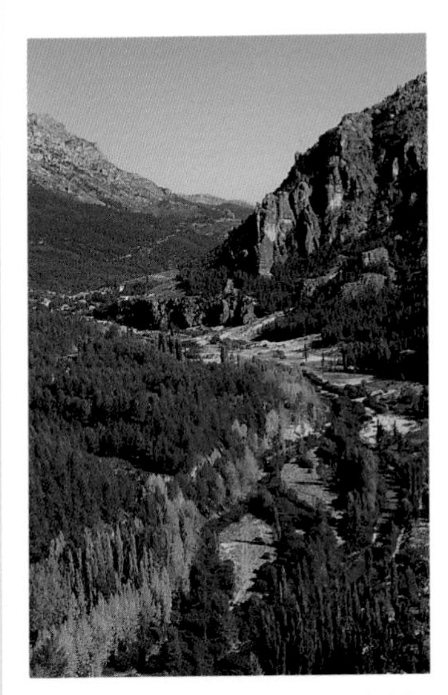

Above: Olive groves that are typical of the north-west of Italy.

Product Box

Le Fasce extra-virgin olive oil

Ceppo Antico extra-virgin olive oil

Taggiasca olives in brine

Vea

Lerida and L'Estornell Organic

Both these distinctively nutty extra-virgin olive oils are made from 100 percent Arbequina olives grown near to the small town of Sarroca de Lleida in the northeastern corner of Spain.

Ninety-four years ago Don Domingo Vea bought the original olive groves and founded the Vea company. His son Don José developed the organic approach which sets the estate apart from other growers in the region. Today, Avelino Vea is in charge and he is a Gran Catador, one of an elite few who are recognized for their tasting abilities.

The hills and valleys of Lérida are ideal for the cultivation of olives for the Mediterranean climate prevents temperatures from rising too high. If you visit the area follow "La Ruta del Aceite," or oil route, along the country roads south of the provincial capital. The harvest starts in November and runs through to the end of January.

The Arbequina olives for Lérida and L'Estornell oils are harvested mainly by hand. The oil is extracted without pressure in the "Sinolea" knife process. The early-picked olives go into Lérida Early Harvest extra-

N *Sarroca de Lleida, Lérida, Spain*

550,000 pounds

★★★ *Excellent*

No

Yes

Above: The Vea factory, Spain.

virgin oil. L'Estornell extra-virgin olive oil (nonorganic) is made up of 60 percent Arbequina from the Vea estates and 40 percent Farga from elsewhere.

The Lérida oil has a warm lemony and almond aroma with a touch of chocolate. The flavor is complex with undertones of grass and toasted nuts which builds to a fiery crescendo. The aftertaste offers an attractive mixture of sweetness and chocolate.

The award-winning L'Estornell Organic oil has an equally interesting aroma with toasted nuts and chocolate. The flavor is equally nutty but softer and less assertive than the Lérida oil. The pepper is light and finish sweetly elegant.

Product Box

Lérida Early Harvest extra-virgin olive oil
Lérida extra-virgin olive oil
L'Estornell Organic extra-virgin olive oil
L'Estornell extra-virgin olive oil

Arbequina, Fraga, and Manzanillo olives in water and salt

Green Bean and Potato Salad

Cook plenty of fresh green beans until just tender or al dente *and steam the same quantity of new potatoes in their skins. Drain and mix the two vegetables and dress with plenty of Lérida or L'Estornell extra-virgin olive oil mixed with a splash of lemon juice, salt, and pepper and very finely chopped herbs. Leave to cool and serve lukewarm or cold. Change the flavor by changing the herbs. Try parsley, chervil, dill or oregano.*

Le Vieux Moulin

For generations the Farnoux family have lived and worked in the old mill at Mirabel aux Baronnies in northern Provence and they are still going strong today, producing medal-winning extra-virgin olive oils.

A hundred and fifty years ago the original Monsieur Farnoux had four sons and under French law the property had to be divided equally between them. The present owner inherited a quarter of the estate from his father and grandfather but has gradually bought up the rest from his many cousins. Both his son and grandson work at the mill and the family connection looks set to continue for many years to come.

The Farnoux family own olive groves in the Beaulieu and Hautes Blanches areas of Mirabel and in Nyons. The countryside in this part of Provence is an attractive mix of vineyards and lavender fields and it is sometimes difficult to find the olive groves which tend to be set on higher terraces.

Mirabel aux Baronnies, La Drome, Provence, France

Not known

★★ *Very good*

No

Yes, by appointment

The groves are planted with the local Tanche olive which is used for both olive oil and table olives. Starting at the end of November, the olives are picked by hand or raked from the trees into nets. The harvest goes on until the end of January. The olives are milled and pressed in the traditional manner at the old mill.

This oil has a very distinctive lemony aroma which is also reminiscent of charcuterie. The flavor is full and eggy with a touch of citrus and not too much pepper. The Vieux Moulin oil won a bronze medal at Nyons in 1991 and a gold medal in 1992.

Pain d'Aubergine

Coarsely chop 1 large eggplant, 1 green pepper and 1 large onion. Add 1 minced clove garlic and cook gently in 3–4 tablespoons of extra-virgin olive oil for half an hour. Purée the mixture in a blender and mix with 3 large beaten eggs and 1 thick slice of crustless bread which has been soaked in milk. Bake in a buttered oven dish in a moderate oven for about 45 minutes. Serve with a tomato sauce made with 1 large can of tomatoes cooked with a little onion, red bell pepper and mixed fresh herbs and puréed.

Product Box

Le Vieux Moulin
extra-virgin olive oil
Olive oil flavored with Provence herbs and with peppers

Black table olives

Tapénade and anchoïade

Rutherford Hill Winery

Lila Jaeger

Lila Jaeger started producing olive oil from old olive trees on the Rutherford Hill Winery estate in California as an "elegant hobby." She now produces about 95 U.S. gallons a year depending on the harvest.

Lila Jaeger decided to press some of the olives she saw going to waste on the Rutherford Estate as a kind of experiment. The olive grove, which is set on a steep hillside, is about 100 years old. The trees are believed to be Nevadillo and Cornezuelo.

The olive trees take up about five acres but so far only about half the grove is in working order. Lila started picking in March but each year the harvest has been brought forward a little with a consequent increase in quality. The olives are now picked as they are turning from green to black. They are knocked from the trees onto nets with poles and processed by the cold centrifugal system.

The oil has a warm aroma of apples and grass. The flavor is nutty with an unusual almost petrol-like tone on the fruit. It has good pepper and an attractive nutty aftertaste.

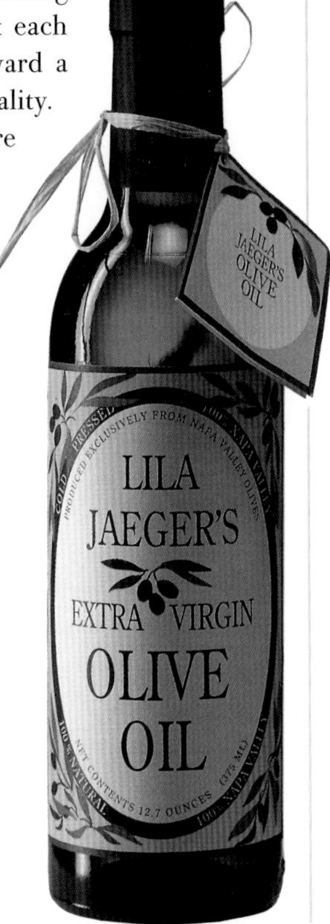

Rutherford, Napa Valley, California, U.S.A.

95 U.S. gallons

★★ *Very good*

Yes

By appointment only

Azienda Agraria

Lungarotti

Olio Extra Vergine dell'Umbria

The electric green color of this first-class Umbrian extra-virgin olive oil is indeed indicative of its distinctive "green" character, but the overall effect is very pleasing.

The Lungarotti family have owned the Azienda Agraria Lungarotti for generations and have always been renowned for their oils and their wines. The olives are grown on hillside groves near Torgiano in the Italian province of Perugia.

Above: The Lungarotti family.

Torgiano, Perugia, Umbria, Italy

4,950 U.S. gallons

★★★ *First-class*

Yes

Yes, by appointment

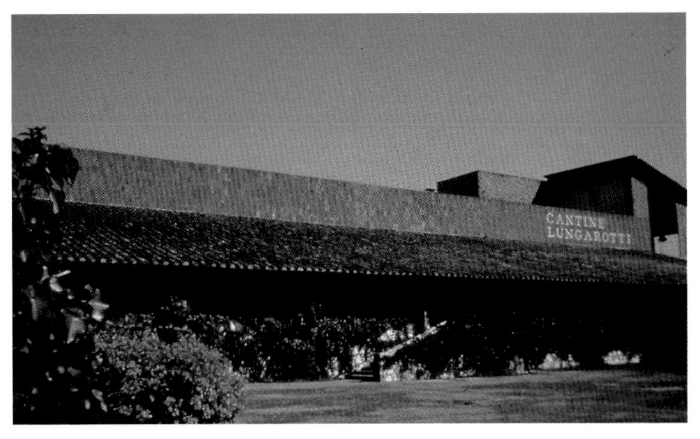

Above: The Lungarotti Winery.

A silver medal winner in the competition, the oil is made up of 30 percent Frantoio, Moraiolo and Leccino with Pendolino making up the rest. The trees are planted in a mixture of traditional and modern groves, the latter with specialized drop-by-drop irrigation systems. The olives are harvested by hand during November and pressed traditionally.

The result is an oil with an amazingly vibrant green color. The aroma is extremely fresh and fruity with grass, salad leaves, and arugula predominating. The taste is aromatically nutty with strong undertones of bitter salad leaves. It packs quite a peppery kick but the aftertaste is surprisingly smooth and sweet.

Umbrian Fava Beans

This dish of fresh fava beans is served in June when the beans ripen. It can also be made with lima beans.

Gently fry 1 chopped onion in 3 tablespoons of extra-virgin Umbrian olive oil with 2 chopped sticks of celery and a small chopped carrot. After two minutes add 1½ cups freshly shelled fava or lima beans, ½ cup chopped Swiss chard or spinach and 3 tablespoons broth or water. Season and cook over low heat for 25 minutes, adding a little more broth if it shows signs of drying up completely. Dress with 3 more tablespoons of olive oil and serve with broiled or roasted meats.

F. Blauel

Mani or Kalamata Gold

Available in conventional and organic form, this fruity Greek extra-virgin olive oil is sold under the brand name Mani in the U.K. but Kalamata Gold or Greek Gold in the U.S.

The reason for this marketing difference is that the name "Mani" was already in use for pure and pomace olive oil in the United States when Mr Blauel started selling his extra-virgin oil.

Mr Blauel started his business in 1980 after discovering what he thought was a hidden treasure on a visit to Mani in Greece. In fact the Mani oil was esteemed throughout the country, though unknown abroad. Nevertheless he was determined to give the oil an international airing.

The Mani extra-virgin olive oils are pressed from Koroneiki olives – not Kalamata olives which are used mainly as table olives. The groves are located on the slopes of the Taygetos mountains which run from Kalamata to Sparta and Gythio, separating the provinces of Messinias and Lakonia.

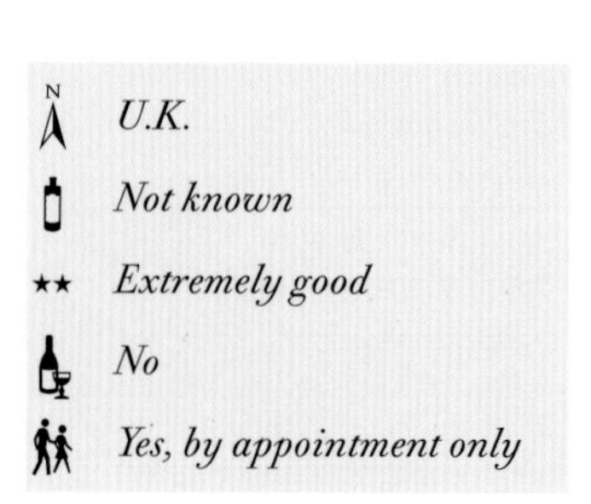

The olives are harvested by hand from November to January. About half of them are milled and pressed in the traditional manner. The rest are pressed by the cold centrifugal process.

The result is a very fruity oil with a flavor very reminiscent of pine resin. The aroma is all freshly mown grass and apple skins and the flavor is full of salad leaves and sweet-sour sorrel with only a touch of pepper.

Cream Horns

Mix together 1 cup all-purpose flour, ¼ cup each white wine and extra-virgin olive oil and 1 teaspoon sugar. Knead well until the mixture no longer sticks to the sides of the bowl. Add a little more flour if necessary. Roll out the dough and cut into strips. Wrap the strips round metal cones and deep fry in extra-virgin olive oil until golden brown. Drain on kitchen paper and remove the cones. Leave to cool and fill with a little preserve and plenty of whipped cream.

Product Box

Mani/Kalamata Gold/Greek Gold extra-virgin olive oil Conventional and organic

Black and green table olives in olive oil

Capers in olive oil

Fratelli

Mantova

Colle Monacesco and Fior Fiore

Set amid the Val di Comino hills at Broccostella in the Frosinone region of Lazio, this company offers a good range of single-estate and blended extra-virgin olive oils.

The company was established in 1905 by Nicandro Mantova in small premises in the town and has grown into a very large modern complex covering 53,000 square feet. Nicandro's tasting expertise has been handed on to his grandson Enzo and great-grandson Adriano who are both involved in the careful blending of most of the oils.

Colle Monacesco, however, is a single-estate oil from a farm in the Colle Monacesco hills near Broccostella. The olives grown here are predominantly Corgnala and Moraiolo. Harvested in November and December, they are hand-picked and pressed traditionally.

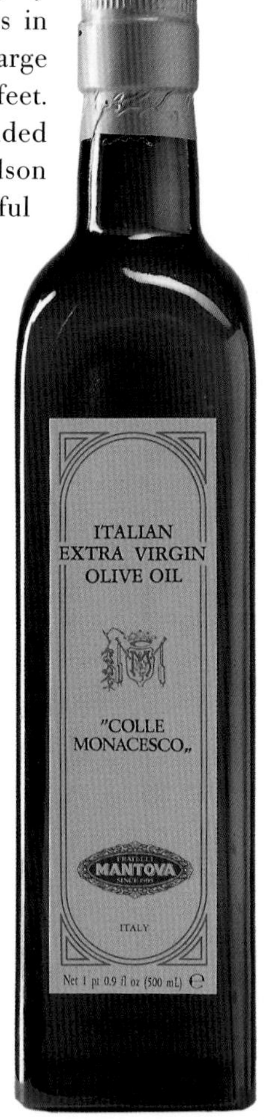

N *Broccostella, Frosinone, Lazio, Italy*

4,750 U.S. gallons single-estate oil, the rest unknown

★★ *Extremely good*

No

Yes, by appointment

The oil has an attractively fresh aroma of grated apples which turns slightly nutty as it ages. The flavor is full of lemons, apples, and nuts with some punchy pepper and a sweet finish.

The olives for the other oils in the Mantova range come from a variety of sources including Greece and Tunisia and are blended to produce good-quality oils. Fior Fiore is an unfiltered oil with a slightly more punchy flavor than Colle Monacesco, though still with the same kind of grated apple aroma and taste. The flavor is more grassy and not quite so elegant.

The Mantova brand extra-virgin oil is a good straightforward oil with an aroma of sweet apples and lemons. Its flavor is smooth and sweet with medium pepper and some grassiness.

Orecchiette with Greens

This recipe uses the ear-shaped pasta with turnip tops but it is very good made with any kind of pasta shapes and well-flavored greens. Cook the oreccheitte in plenty of boiling salted water until just tender or al dente. *Wash a large bunch of spinach, Swiss chard or pak-choi. Place in a pan and allow to wilt over low heat. When the pasta is cooked, drain well and toss in plenty of extra-virgin olive oil. Drain the greens and add to the pasta. Toss again and serve.*

Product Box

Colle Monacesco extra-virgin olive oil

Fior Fiore extra-virgin olive oil

Fior Fiore Novello extra-virgin olive oil

Antica Abbazia extra-virgin olive oil

Mantova extra-virgin olive oil

Grand Aroma flavored olive oils

Mas de la Dame les Baux

Immortalized in the paintings of Van Gogh, this seventeenth-century domaine produces organic extra-virgin olive oil from the local Picholine olive.

The property is situated at the foot of the dramatic hill fortress of les Baux de Provence and with its 60 acres of olive groves and 130 acres of vineyard it is the largest oil and wine domaine in the region.

Both Picholine and Grossane olives are grown on the farm and as well as being pressed for oil they are cured as table olives. The extra-virgin olive oil has an unusual aroma of nuts and chocolate with a touch of ham or charcuterie. The flavor is similar, with meaty olives and toasted nuts. There is just a little pepper on the finish. Serve on well-flavored salads.

Product Box

Mas de la Dame extra-virgin olive oil

Black and green table olives
Olives Cassées au Fenouil

Tapénade

Les Baux de Provence, Arles, Provence, France

Not known

★★ *Very good*

Yes

Yes

Masseria

A consortium of family-run farms in the Aquaviva delle Fonti region of Apulia near Bari produce this smoothly sweet organic extra-virgin olive oil.

Until 1990 most of these farmers sold their oil to large blenders and bottlers. Today they have created their own organization where only the very best organically-grown olives are allowed into the blend which is 60 percent Coratina and 40 percent Ogliarola.

It has a light fruity aroma with a mix of apples, nuts, and grass. The flavor is reminiscent of freshly grated apples, then a strong nutty flavor, sometimes with a touch of chocolate and fiery pepper. The aftertaste is pleasantly fruity.

Product Box

Extra-virgin olive oil

Vegetables in oil, including artichokes, sun-dried eggplants, and broiled onions

Sauces including Southern Broccoli, Pizzaiola, and Bell Pepper

N *Aquaviva delle Fonti, Bari, Apulia, Italy*

1,600 U.S. gallons

★★★ *First-class*

No

Yes, by appointment only

The Merchant Gourmet

This U.K. importer sources good-quality extra-virgin olive oils, often from single estates, in the countries of origin and sells them under its own brand name.

There will eventually be six oils in the Merchant Gourmet range but only three are currently available. They are from Spain, Greece, and Italy.

The Merchant Gourmet Spanish extra-virgin olive oil is a single-estate oil made from 100 percent Picual olives in Córdoba. The olives are hand-picked and the oil extracted without pressure in the "Sinolea" knife method. The oil has a sweet lemony aroma with melon tones. The taste is slightly peppery with an even stronger flavor of Cantaloupe melons. This is one of the best oils in the Merchant Gourmet range.

The Greek extra-virgin olive oil comes from a family farm set in mountainous countryside in Kydonia in Crete. In the past their oil was sent to the big local cooperatives and this is their first attempt to sell the oil direct to the customer through the Merchant Gourmet.

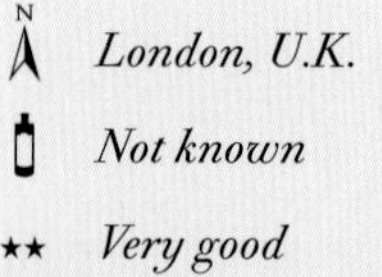

The oil is made from 100 percent Koroneiki olives which are traditionally harvested and pressed. It has a light aroma of warm hay and cut grasses. The flavor is extremely grassy with arugula and sorrel leaves giving a distinctive bitterness to the aftertaste. There is plenty of pepper. This is a definite oil which is also attractively rounded.

The Italian extra-virgin olive oil comes from the firm of Crespi e Figlia in Liguria. Forty percent of the oil is pressed from local Taggiasca olives and the rest is pressed from Leccino and Oliarola or Frantoio olives from other parts of Italy. The resulting oil has a pleasant lemony aroma and a lightly grassy flavor. The pepper is light.

Spaghettini with Black Olive Paste

Cook the spaghettini until just tender or al dente. *Seed and finely chop 2 fresh chili peppers and cook in a little oil with 2 sprigs rosemary, 2 sprigs sage, 2 minced cloves garlic, and some dried oregano. After a couple of minutes add the contents of a small jar of black olive paste and dilute with some of the cooking water from the pasta. Season to taste. Toss the cooked pasta in this sauce.*

Product Box

Spanish extra-virgin olive oil
Greek extra-virgin olive oil
Italian extra-virgin olive oil

Herbs in olive oil

Olives and olive pastes

Vegetable sauces

Meridian Organic

This company specializes in organic foods and their extra-virgin olive oil is no exception.

This is a Spanish oil which comes from organic producers near Córdoba in Andalusia. It is certified by the U.K. Soil Association and by the Spanish Comite Territorial Andaluz de la Agricultura Ecologica.

Meridian extra-virgin olive oil is pressed from 100 percent Hojiblanca olives which are grown without fertilizers or pesticides. The fruit is harvested by hand and processed by the cold centrifugal system.

The oil has an aromatic lemony aroma, sometimes with a touch of tomato coulis. The flavor is smooth and lemony with a medium peppery finish.

Marinated Monkfish Fillets

Slice the monkfish and cook in boiling salted water. Meanwhile mash 1 hard-boiled egg with 4 anchovy fillets and 10 chopped capers. Add a dash of tarragon vinegar and gradually add about 4–5 tablespoons of extra-virgin olive oil. Stir in some freshly chopped parsley. Drain the fish and spoon the egg and olive oil mixture over the top of each slice of fish. Leave to cool and serve lightly chilled.

Corwen, Clwyd, U.K.

Not known

★★ *Very good*

No

No

Frantoio Valtenesi

Mirum

The beautiful Lake Garda is the backdrop for the hillside estate where the olives are produced and are pressed to make this extremely attractive extra-virgin olive oil.

The Caldera family has been running the Frantoio Valtenesi since the 1960s, and are members of the Mastri Oleari.

Mirum is a single-estate oil which is made up of a blend of 70 percent Leccino and 30 percent Casaliva olives.

The oil has a gentle aroma of apples and almonds. The flavor is sweet and fruity with a delicate touch of toasted nuts. There is just a light feathering of pepper leading to an extremely pleasant aftertaste.

Product Box

Mirum extra-virgin olive oil
Arzane extra-virgin olive oil
Gardolio extra-virgin olive oil

Table olives in brine, pitted and unpitted

Olive paste

Arzane di Polpenazza, Brescia, Lombardy, Italy

30,000-40,000 bottles in total

★★★ *First-class*

No

Yes

Fattoria
Montevertine

This small wine estate near Radda-in-Chianti produces first-class extra-virgin olive oil from an ancient grove which has been extended with new trees.

The Montevertine estate is a typical Italian family business. It was bought by Sergio Manetti in 1967 and he now runs it with the help of his son-in-law and three children and the local Bruno Bini family.

Three olive varieties are used in equal proportions. They are Correggiolo, Moraiolo, and Leccino. They are harvested by hand, starting at the beginning of December. Once picked they are milled and pressed in the traditional manner.

The oil has a strong aroma of fresh olives, cut grass, and grated apple skins. The taste is bittersweet, with nutty flavors amid fruitiness. There is medium pepper with a sweet aftertaste.

As well as tending the vines and the olive groves Sergio Manetti also collects agricultural tools and equipment. Over the last 40 years he has built up an impressive collection. Everything is now housed in a small museum which is open all the year round. It is well worth a visit.

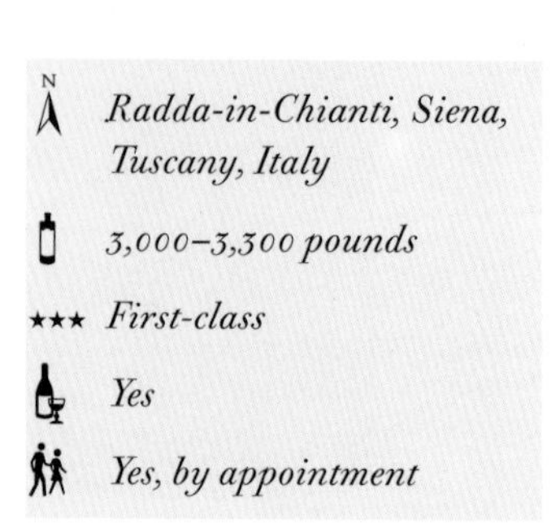

Radda-in-Chianti, Siena, Tuscany, Italy

3,000–3,300 pounds

★★★ *First-class*

Yes

Yes, by appointment

Pipolo Silvio

NATIVO

This delicately fresh extra-virgin olive oil was chosen to represent Campania in the national "Ercole Olivario" last year and won third prize.

This small family-run estate is situated in a remote part of central Campania between Laurino and Valle dell'Angelo. The groves are planted with Biancolilla Del Cervati olives and these alone are used for the Nativo oil. The company also sells two other oils which are pressed from olives grown on nearby estates and these include Frantoio and Rotondella as well as the Biancolilla olives.

The Nativo oil is produced by the *percolamento naturale a freddo* cold dropping process and is decanted three times before bottling to order. It has an aroma of freshly grated apples and a delicate flavor of apples and nuts with a hint of chocolate. There is medium pepper which fades to a smooth nutty aftertaste.

PRODUCT BOX

Nativo extra-virgin olive oil

Viride extra-virgin olive oil

Valle dell' Angelo extra-virgin olive oil

Laurino, Campania, Italy

2,800 U.S. gallons

★★ *Extremely good*

No

Yes

NUÑEZ DE PRADO

Seven generations of experience go into the unpressed "flower" of the oil for this wonderful, prize-winning, organic extra-virgin olive oil from Andalusia.

The Santa Lucia mill, where the Nuñez de Prado family crush and press their olives, is situated in the heart of Baena. This Denomination of Origin region produces some of Spain's sweetest olive oils. The Nuñez de Prado family have been producing olive oil here since 1795.

The company now has 1,480 acres of olive groves on four estates. They are planted with Picudo, Hojiblanca, and Picual varieties. There are still some old three to four trunk trees on the estate but most are now single-trunk trees planted much closer than in the past. Cultivation is fully organic with no artificial fertilizers or pesticides.

The fruit is hand-picked and washed in the groves before being transported to the Santa Lucia mill. Here the olives are stone-crushed in the traditional manner. The paste then undergoes a partial extraction to produce the "Yema" yolk or "flor" flower of the oil. The paste is put into drums which slowly revolve.

Santa Lucia, Baena, Andalusia, Spain

100 tons by partial extraction and 100 tons by traditional pressing

★★★ *Excellent*

No

Yes, mornings only

Picada for Scallops or Shrimp in Tomato Sauce

Make a quick sauce by frying some shallots and garlic in extra-virgin olive oil with ripe, roughly chopped fresh tomatoes. Add tomato purée and seasonings to taste. Next fry a slice of white bread in more extra-virgin olive oil until lightly browned. Chop and process the fried bread in a blender with 2 tablespoons ground almonds, half a red bell pepper, a little freshly chopped parsley and dill, and a pinch of thyme. This is the picada. Keep to one side while you quickly stir-fry fresh scallops or shrimp for 2–3 minutes until done. Stir the picada into the tomato sauce and serve spooned over the cooked shellfish.

Gradually the free-run oil seeps out. It takes 24 pounds of paste to produce 1 quart of oil by this method.

The oil obtained in this way is bottled in a distinctive square bottle which bears a numbered label and the corks are sealed with sealing wax. This wonderful extra-virgin olive oil has a very consistent aroma and flavor and varies very little from year to year. The aroma is full of fruit with lemons, apples, melons, passion fruit, and bananas jostling for place. The taste is just as good, with the same exotically fruity flavor. There is only a touch of pepper with a lovely smooth, sweet aftertaste.

Salmorejo

Make the garnish for this soup first. Simply hard-boil an egg and fry a large slice of bread in extra-virgin olive oil. When they are cold dice the egg and the bread and add about 3 tablespoons chopped serano ham to the mix. Reserve. Next process 1 pound fresh breadcrumbs moistened with water in a blender or food processor with 1 large, seeded green pepper, ½ pound of ripe tomatoes, 3 cloves garlic and 1 cup Nunez de Prado extra-virgin olive oil. Add a splash of sherry vinegar to taste. Serve with the garnish sprinkled over the top.

Oakville Grocery Co.

This enterprising store in St. Helena, California is expanding its range of extra-virgin olive oils at a rapid rate in line with the growing interest in olive oil in the U.S.

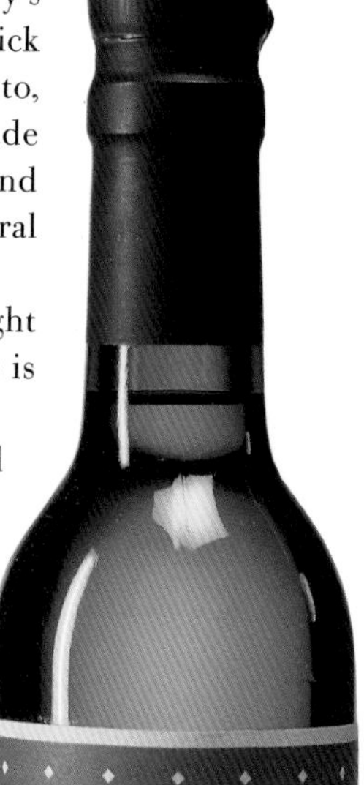

At the moment Oakville Grocery's olive oils are produced by Nick Sciabica and Sons at their mill in Modesto, California. The main Oakwood oil is made up of a special blend of Mission and Manzanillo olives grown in the Central Valley of California.

The extra-virgin olive oil has a light aroma of apples and nuts. The flavor too is light and nutty with grassy overtones.

Oakville Grocery also market a limited production of olive oil from olives grown in the Napa Valley. Last year the production was only 50 cases. They hope to expand the range of these special edition Californian olive oils in the future.

- *St. Helena, California, U.S.A.*
- *Not known*
- ★★ *Very good when fresh*
- *No*
- *Yes*

Olis di Catalunya
OLEASTRUM

Fifty-eight cooperative producers in the Spanish D.O.C. areas of Siruana and Les Garrigues have banded together to market their extra-virgin olive oil under the Oleastrum brand name.

This Catalan consortium was set up in 1993 and now has around 3,200 participating farmers. The Siruana D.O.C. area is in Tarragona. It runs inland from the coast south of Barcelona to La Palma de Ebro and Margalef de Montsant. Les Garrigues is set further inland south of the city of Lérida. This D.O.C. area used to be known as Borgas Blancas.

To qualify for the D.O.C. the oils must be pressed solely from the Arbequina olive. Harvesting is carried out by hand and the olives are processed by the continuous cold centrifugal system. All bottles are certified and numbered.

Oleastrum extra-virgin olive oil has an attractive aroma of freshly grated apples and lemons with a touch of almonds. The flavor is light and fruity with roasted nuts. There is light to medium pepper but the overall effect is very sweet. This is a versatile oil.

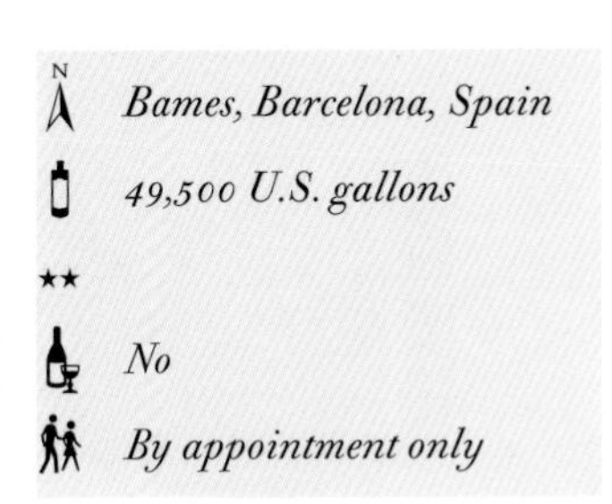
Bames, Barcelona, Spain

49,500 U.S. gallons

★★

No

By appointment only

Podere di Ursini

Opera Mastra and Gentile di Chieti

This family-run estate on the Adriatic coast of Italy produces stunningly good extra-virgin olive oils which rival the best that any other region can produce.

Guiseppe Ursini carries on the family tradition of meticulous care for quality in his management of this abundant estate just south of Pescara in the Abruzzi. The Ursini have been producing oil here for four generations. They are members of the prestigious Mastri Oleari.

The olive groves, on a marvelous promontory overlooking the Gulf of Venus, are planted with Gentile olives, known as the king of olives in this part of the world, as well as with Leccino, Nebbio, and Cucco varieties.

All the olives are picked by hand. The harvest starts at the beginning of November and is complete by the end of December. Processing is partly by traditional mill and hydraulic press and partly by the more modern centrifugal process.

The Opera Mastra oil in particular demonstrates the unique flavor of herbs, wild flowers, and freshly mown hay that is characteristic of oil made from the Gentile olive. The scent is fresh and fruity with apples and herbs.

San Giovanni in Venere, Fossacesia, Abruzzi, Italy

9,500 U.S. gallons

★★★ *Excellent*

No

Yes, by appointment

The taste is reminiscent of fresh cilantro, tarragon, and spices with salad leaves and pepper. It needs nothing more to make a perfect salad dressing.

Gentile di Chieti is equally sweet but the flavors are softer and less herbaceous with more nutty almond flavors. This is an elegant oil with light to medium pepper.

Guiseppe Ursini also uses some of his olives to make Agrumolio, a fruit-flavored oil with a great depth of flavor which comes from milling oranges and lemons with the olives. This oil is best used on its own as a condiment on lightly cooked vegetables and fish.

Roasted Fish with Orange-Flavored Olives

Roast large steaks or a whole fish in olive oil and a little white wine in a covered dish in the oven. Just before cooking sprinkle a dozen or so orange-flavored black olives and some of the orange peel around the fish. Season with freshly ground black pepper.

Product Box

Opera Mastra extra-virgin olive oil
Gentile di Chieti extra-virgin olive oil
Agrumolio orange- and lemon-flavored olive oil

Black olives with orange peel
Cracked olives with fennel or chili peppers

Black olive paste
Green olive paste

Vegetables in olive oil

Tenuta dell' Ornellaia

Tenuta dell'Ornellaia is run by the Marchese Lodovico Antinori quite independently of the rest of the family estates. The farm produces excellent wine and first-class extra-virgin olive oil.

Situated not far from the sea in a picturesque corner of Tuscany 60 miles south of Florence the property supports 1,700 mature olive trees. The groves are planted on elevated hillsides at the fringe of the Mediterranean marsh, between Bolgheri and Castagneto Carducci. The estate borders the Tenuta San Guido farm owned by the Marchese's uncle.

The trees are planted on poor soil between the marsh scrub and cork trees and are exposed to the sea winds. As a result the olives tend to be very small but they have a very intense flavor which gives the oil its strong character. They are typical Tuscan olives – Frantoio, Moraiolo, and Leccino. Picking starts quite early on 25 October and continues until the end of November. The olives are traditionally milled.

Tenuta dell'Ornellaia extra-virgin olive oil has a pleasing aroma of grass, apples, and salad leaves. The flavor is equally fresh and fruity with a touch of bitterness and medium pepper. The aftertaste is extremely smooth and attractively leafy.

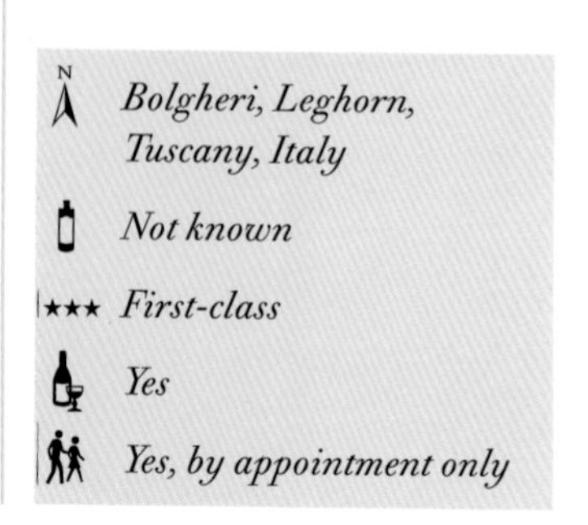

N *Bolgheri, Leghorn, Tuscany, Italy*

Not known

★★★ *First-class*

Yes

Yes, by appointment only

Pasolini Dall'Onda

Laudemio

A member of the Italian Laudemio marketing consortium, Passolini dall'Onda single-estate extra-virgin olive oil is produced on a Tuscan estate with a history stretching back to the fourteenth century.

The olive groves, which are the personal property of the Contessa Pasolini dall'Onda, are adjacent to the ancient Palazzo dei Pandolfini in Barberino Val d'Elsa in Tuscany. The village and its surrounds, which lie about halfway between Florence and Siena, belonged to the Pandolfini family until 1573 when they were inherited by the Pasolini dall'Onda.

The estate produces first-class Chianti wines and two olive oils. One is sold under the Laudemio banner – look for the *raccolta* or harvest date on the label – and the other under the estate's own name. Both are produced from a blend of 50 percent Frantoiano, 30 percent Morellino, and 20 percent Leccino, but the Laudemio oil comes from selected olives.

The groves are planted on land which is particularly suited to the cultivation of olives both in terms of the lie of the land and its soil. The olives are hand-picked and processed in the traditional manner.

Barberino Val d'Elsa, Florence, Tuscany, Italy

2,600–2,900 pounds

★★ *Extremely good*

Yes

By appointment

The Laudemio oil has a strong aroma of freshly-cut grass and grated apple skins with almonds. It has a similar taste with bitter undertones and plenty of pepper. The aftertaste is smooth and nutty. The second oil is less distinctive with a light and fruity aroma of warm olives, apples, and almonds. The taste is sweet and nutty with a strong peppery aftertaste.

Salmon with Barberino Sauce

Place four salmon steaks in a deep skillet. Add 1 sliced onion and carrot and cover with water. Bring to a boil and boil for 1 minute. Turn off the heat and leave the salmon to stand for 15 minutes. Drain off 1 cup of the fish stock and boil with a bunch of fresh herbs (dill, basil, parsley, bay leaf) for 5 minutes. Remove the herbs. Melt a small piece of butter in a small pan and stir in 2 level tablespoons all-purpose flour. Add the broth and bring to a boil again. Transfer to a blender and process, gradually adding $1/4$–$1/3$ *cups extra-virgin olive oil. Pour onto individual plates and serve the salmon steaks on the top.*

Phoenician Olive Oil

Manzanillo

Arizona is the unlikely base for this relatively new company which presses and packages extra-virgin olive oil from olives grown on the Gila River Indian Community Reservation in the North Sonoran Desert, 20 miles south of Phoenix.

The project started in the late 1980s when the Indian Community agreed to plant 200 Manzanillo olive trees a year as part of their farming program. The first olives came on stream in sufficient quantities to press in 1994.

The olives are harvested by hand between September and December. They then go to the Phoenician Olive Oil Company's mill on site where the oil is extracted using a cold centrifugal process.

When it is fresh the oil has a very attractive grassy, grated-apple aroma. The flavor, too, is grassy with a touch of woody bitterness on the aftertaste. It is not very peppery.

This company also produces oil from Pendolino olives at Caborca in Mexico. This oil is not as good as the Gila River oil.

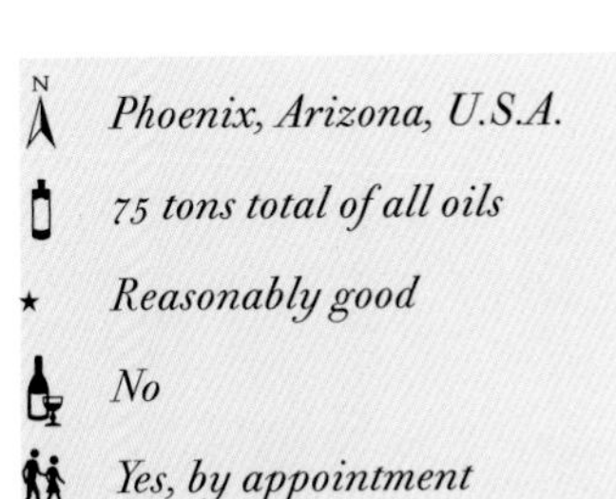

Podere Cogno

Trees ranging in age from six to 600 years provide the olives for this fine extra-virgin olive oil produced by an English family living in Tuscany.

Abandoned in the general Italian flight from the land in the early 1960s, this estate was bought by Alexander Oldbrook and his wife Eva. The original olive grove faces west–southwest. It contains 1,600 trees some of which are original, some planted when the Oldbrooks arrived and some planted to replace the devastation after the 1985 frosts.

The estate now has two further olive groves, one planted six years ago on an old south-facing terrace, containing 1,000 trees and a brand new one containing 400 trees planted on newly-cleared terraces. The olives from all the groves are harvested by hand from mid November before being processed by the cold centrifugal system. The oil is available either filtered or unfiltered.

The mix of olives in the oil is 80 percent Frantoio and ten percent each of Moraiolo and Leccino. The resulting extra-virgin oil has a rich fruity aroma with bitter sorrel leaves and cut grass. As it ages it takes on a nutty, chocolaty aroma with vanilla. The flavor of the young oil is full of bitter leaves and cut hay with good pepper and a sweetish aftertaste. This softens over time to a nutty almost cakey flavor.

N *Castellina-in-Chianti, Siena, Tuscany, Italy*

4,400 pounds

★★ *Extremely good*

No

By appointment only

Poggio Lamentano

The Zyw family had to learn about olives and olive oil from their neighbors when they first bought this Tuscan estate in 1961. Today they produce an extra-virgin oil which is applauded far beyond their own horizons.

In Renaissance Italy Leonardo da Vinci imagined a way of life based on creativity and the earthy work of a farmer and this vision has turned to reality at Poggio Lamentano where olive-growing is combined with painting.

Aleksander Zyw, painter and war artist, bought the estate in 1961. His son Michael, also a painter, joined him on the estate with

Green Carpaccio

This recipe comes from the restaurant Il Bugaile in Guardistallo. Thinly slice one or two bulbs of fennel. Arrange on individual plates and top with thinly sliced raw mushrooms and fresh Parmesan flakes. Add a few drops of lemon juice, salt and pepper, and plenty of Poggio Lamentano extra-virgin olive oil.

 *Castagneto Carducci, Leghorn, Tuscany, Italy*

8,800–13,200 pounds

★★★ *Excellent*

 *No*

 *Yes, by appointment only*

his family in the 1970s. Except for olive-picking, which needs more hands, the family do all the work on the farm themselves.

Over the years they have developed an award-winning oil which caught the eye of the English cookery writer Elizabeth David. She helped to pioneer the sales of olive oil in the U.K. and sold the oil in her shop in London.

The estate is situated not far from the Tuscan coast south of Leghorn in an area with few tourists. The olive groves are set on a rocky hillside and some of the trees have been there for 200 years or more. The fruit from these trees is mixed with the crop from a young plantation which is only about 10–12 years old.

The olives are harvested by hand from late October to early December. The process is both traditional and centrifugal and the oil is unfiltered. The blend of olives is 60 percent Moraiolo, 30 percent Frantoio, and five percent each Leccino and Pendolino.

This results in a very fresh olive oil, with an aroma of apple skins and cut grass. The taste is lush and interestingly grassy with nutty overtones. It is an assertive but not aggressive olive oil with medium pepper.

Above: Olive picking.

Ponder Estate

This small estate is pioneering the development of olive varieties suited to the climate and soils of both Australia and New Zealand.

Mike Ponder and his wife Diane bought the land in Blenheim, Marlborough county, in 1987. They started the groves by importing 200 Barnea olive trees from Israel. This was followed by a further 200 the next year. They now have 3,000 trees and are already producing around 528 U.S. gallons of extra-virgin olive oil.

The Ponders believe that New Zealand could become an important producer of high-quality olive oil and they are well on the way to making this prediction come true.

Their oil is currently pressed from 90 percent Barnea olives and ten percent Manzanillo olives, but this could change as 15 different varieties have been planted in the Ponder estate groves. The trees are planted on flat, well-drained shingle soils.

The olives are raked onto cloth nets and pressed in the traditional manner. The resulting extra-virgin olive oil has a wonderfully strong aroma of freshly grated apple skins. Apples continue in the taste but with an interesting mixture of dried hay, bitter salad leaves, and a hint of lightly toasted almonds or chocolate. The oil is full-flavored but not sweet and there is plenty of pepper with a smooth nutty finish.

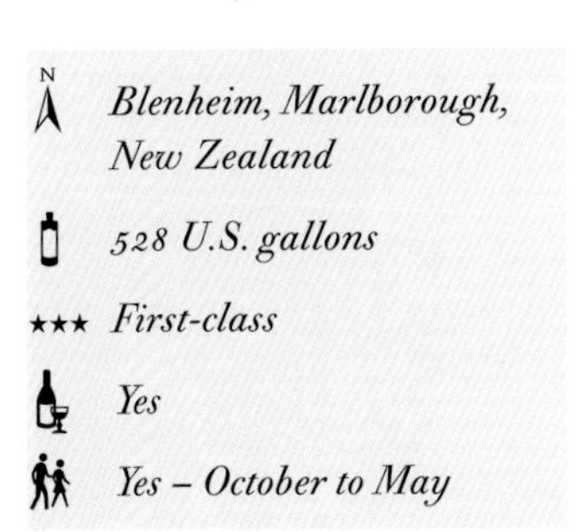

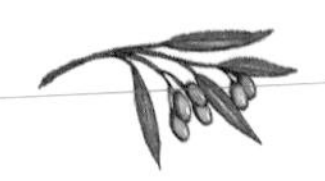

Possenti Castelli

One of the present olive groves at Possenti Castelli was planted around 1640. It replaced a previous grove which had been severely damaged by frost. The hazards of growing olives do not change!

The Possenti Castelli estate is set amid the Umbrian hills near Rocca San Zenone, just east of Terni. The trees in the older groves have been there for so long that they have grown particularly large and tall. They offer an unusual contrast to the newer groves, planted in the modern pattern.

The main varieties are Frantoio and Moraiolo with a small amount of Leccino. The harvest runs from November to January. The olives are picked by hand and pressed traditionally.

The Possenti Castelli extra-virgin oil has a strongly fruity aroma with plenty of cut grass and salad leaves. The flavor is sweet and fulsome with green salad leaves and medium pepper. It is an elegant oil.

Product Box

Extra-virgin olive oil
Flavored extra-virgin olive oils

Olive Pastes

Midnight Sauce
Devil's Sauce

- *Rocca San Zenone, Terni, Umbria, Italy*
- *61,600 pounds*
- ★★★ *First-class*
- *No*
- *Yes, by appointment*

Antica Azienda

Raineri

Prela, Riviera Ligure di Ponente, and Selezione Primavera Mosto

This well-known Ligurian company offers extra-virgin olive oil from their own farm in the ancient village of Praelo as well as blended oils from the whole of the Ponente region.

The Raineri farm lies in the heart of Ligurian olive territory in a valley named La Valletta, or "little valley," about 12 miles inland from the sea. It is protected to the north by the Alpes Maritimes and the olive trees thrive in the chalky rock of the region. The Taggiasca olive is king here and only this olive variety is used in the Raineri oils.

Prela extra-virgin olive oil, packed in an elegantly tall and slim bottle, is the Raineri family estate oil. The flavor is smooth and sweet with a mixture of fruit and nuts. The pepper is light.

You cannot miss Raineri Riviera Liguria di Ponente extra-virgin olive oil – it is completely wrapped in gold foil. The oil has a wonderfully fresh and lemony aroma with apples and almonds. The flavor is smooth and sweet with a mixture of almonds and lemons. The light peppery taste builds up quite strongly but with a smooth aftertaste.

Conservas

Rainha Santa and Triunfo

This company is run by an English couple who source their extra-virgin olive oils from the Alentejo region of eastern Portugal.

Mervyn and Katherine Clements bought Conservas Rainha Santa in 1990. The Company specializes in olives and olive oil and the unusual but delicious bottled Elvas plums. The oil for Rainha Santa extra-virgin olive oil comes from 100 percent Galega olives grown by selected farmers in and around Estremoz in the Alentejo but the net is cast rather more widely for the Triunfo oil which comes from olives grown throughout the area of Estremoz, Borba, and Elvas.

The groves are mainly organic. The olives, picked in late November to late December, are milled and pressed traditionally. The Rainha Santa extra-virgin olive oil has a light lemony aroma with earthy herbs. The flavor is distinctive with aromatic flavors of dried herbs, pine needles, and dried ginger.

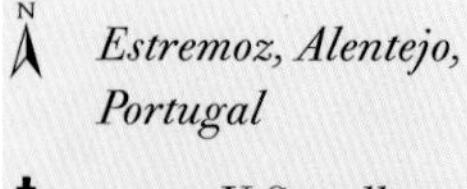

Estremoz, Alentejo, Portugal

5,300 U.S. gallons

★★ *Very good*

No

Yes

Azienda Agricola
RAVIDA

This delicious organic extra-virgin olive oil has placed Sicily firmly on the quality olive oil map. In 1993 Ravida became the first Sicilian oil to win a national competition.

The Ravida family, now based partly in Rome and partly in Sicily, have been associated with the beautiful La Gurra estate since the mid eighteenth century. Situated in the southern part of the island near Menfi, the estate is dominated by the sixteenth-century farmhouse which overlooks the Mediterranean sea facing the Greek temples of Selinunte.

There are 130 acres of olive groves and 120 acres of vineyard as well as lemon groves and fields of artichokes and wheat. The farm is run on organic principles and no artificial fertilizers or pesticides are used.

The olive groves are planted with Cerasuola and Biancolilla and a small number of Nocellara del Belice olives. The latter variety can also be used to make table olives. The harvest starts in the first week in November when the Cerasuola turns from green to violet and while the other two varieties are still unripe.

The olives are picked by hand with the help of vibrating combs which allow the

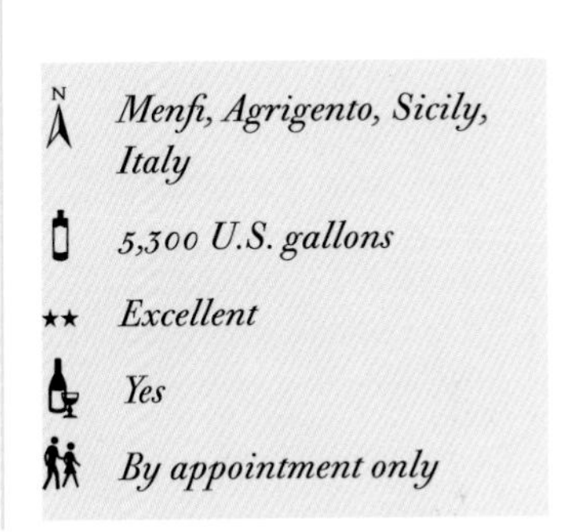

Above: Ravida family home and estate.

olives to fall into nets placed under the trees. They are then processed by the cold centrifugal system within eight hours of being picked. The oil is unfiltered.

This distinctive oil has an intense aroma of tart apples and cut salad leaves with just a touch of tomato skins. After a while the aroma softens and the tomato skins increase. The flavor is equally definite with olives, apples, and grass moving to a bitter pepperiness. The aftertaste is attractively smooth with a flavor reminiscent of sweet-sour sorrel leaves.

Nicolo Ravida, the present owner, is deeply committed to teaching people how to enjoy olive oil to the full and the estate is used for culinary events and cookery courses with top chefs.

Winter Salad

Cut two bulbs of raw fennel into quarters and place in cold water for 10–15 minutes. Rinse, drain, and slice very thinly. Slice 1 orange and 1 apple. Mix all three ingredients in a bowl with bitter salad leaves such as arugula and black olives. Season with salt and plenty of Sicilian extra-virgin olive oil.

Conservas

Rocca Delle Macie

This very well-known wine estate in the Chianti Classico area of central Tuscany also makes a first-class extra-virgin olive oil.

Situated on a Burton Anderson wine route at Castellina-in-Chianti, halfway between Florence and Siena, this estate is part of a large corporation headed by film producer Italo Zingarelli.

The extra-virgin olive oil is made from typical Tuscan olive varieties which are hand picked in October. The resulting oil has a wonderful aroma and flavor of freshly grated apple skins. There is also an interesting hint of grass and bitter salad leaves with good pepper. The aftertaste is pleasantly sweet.

Product Box

Triunfo extra-virgin olive oil

Green olives in oil with mixed herbs
Black Galega olives in oil with mixed herbs
Mixed Beira Baixa
Green Elvas tapénade
Black Galega tapénade

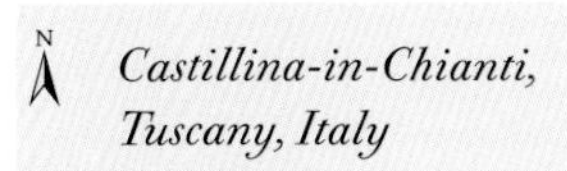

Castillina-in-Chianti, Tuscany, Italy

Not known

★★★ *First class*

Yes

Not known

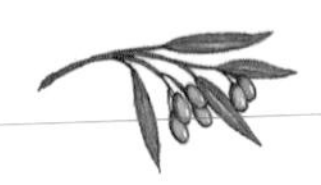

Fattoria

Roi

The olives for this extra-virgin olive oil come from 100- to 600-year-old trees planted in the hills behind Taggia in western Liguria.

Fattoria Roi was founded in 1900. It has changed location several times but is now settled in new premises at Badalucco on the banks of the Argentina river which used to power the old mill. Pippo Roi, grandson of the founder, has modernized operations, installed electricity, and reorganized the olive groves.

The olives come mainly from old groves planted on terraces up the steep valley sides between 200 and 1,400 feet. Taggiasca is the sole variety. The harvest lasts from November to March. Once picked the olives are milled and pressed in the traditional manner. A percentage of the crop is used for the oils but the rest is used for table olives and olive pastes.

Roi extra-virgin olive oil has a very attractive light almond aroma with apples and a touch of grassiness. The flavor is sweet and nutty with light pepper and a little tart bitterness on the aftertaste. This is a very smooth, delicately complex and versatile oil.

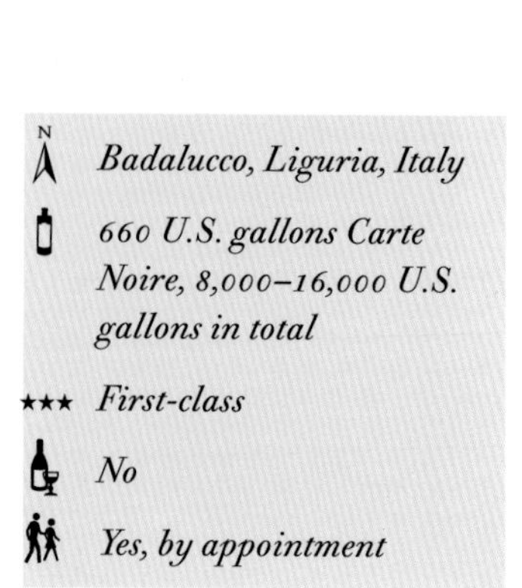
Badalucco, Liguria, Italy

660 U.S. gallons Carte Noire, 8,000–16,000 U.S. gallons in total

★★★ *First-class*

No

Yes, by appointment

Olex

Romanico

This is the brand name for the oils of an old-established and respected Catalan cooperative.

The Olex cooperative was set up about fifty years ago and it now works with over 6,000 growers in 42 villages in the region of Lerida. Five years ago 25 to 30 farmers in one of these villages decided to go organic and stopped the used of pesticides and artificial fertilizers.

Both oils are pressed from the Arbequina olive. Romanico extra-virgin olive oil has an aroma of melons, lemons, and almonds. The flavor is sweetly nutty with cut leaves. There is a surprising crescendo of pepper, but the aftertaste is smooth and fruity.

The organic extra-virgin olive oil has much more tropical fruit in the aroma with strong hints of passion fruit and lots of melon. The flavor is even sweeter than the ordinary extra-virgin oil with more tropical fruits and eggy lemons and a good peppery finish. After a little while the oil takes on even more complex flavors with touch of coffee or chocolate.

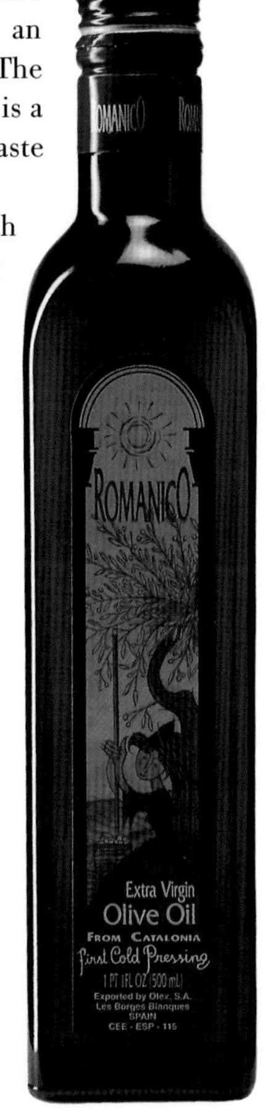

N *Les Borges Blancues, Las Garrigues, Lerida, Catalonia, Spain*

Not known

★★ *Extremely good – Romanico*

★★★ *First class – Romanico organic*

No

No

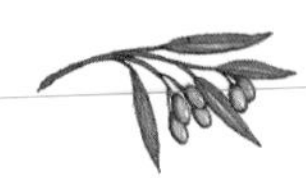

Sadeg Organic

An olive grove which is more than 100 years old, organic status, and a dedication to making the best-quality premium extra-virgin oil is the unique combination behind this sweetly nutty oil.

Situated at Palermo in California, the Sadeg ranch is about 60 miles north of Sacramento. It was named after Palermo in Sicily because it was thought to have the same kind of countryside and a similar climate. It is in fact very hot and dry in summer and cool and damp in the winter. The region is one of mixed farming with a rice farm, dry wheat grower, and cattle ranch nearby.

The olive grove is planted with Manzanillo as well as Mission Olives but only the Mission olives are pressed for the oil. The grove is certified by the Organic Crop Improvement Association (O.C.I.A.) and only biological and natural methods of fertilization and pest control are used.

The olives are picked when they are fully ripe and black in color. This is usually about December through January. The olives are processed by the centrifugal method. The relatively late harvest results in an oil which is very smooth and buttery.

The oil has a full lemony and apple aroma and a warm flavor of ripe olives and almonds. It does not have very much pepper and its lack of harsh bite makes it ideal for those who prefer a milder oil.

N *Palermo, California, U.S.A.*

1,000–1,300 U.S. gallons increasing

★★ *Very good*

No

Yes, by appointment only

Azienda Agricola

S. Cristina

The Zenato family produces a sweetly attractive, fully organic extra-virgin olive oil from their groves on the eastern shores of Lake Garda.

Azienda Agricola S. Cristina is situated at San Benedetto di Lugana near Peschiera del Garda. As well as producing olive oil, the Zenato family have been producing fine wines from their vineyards here since the 1930s.

The olives for the S. Cristina extra-virgin olive oil are a blend of Frantoio, Leccino and Casaliva, but the predominant variety is Frantoio. The olives are processed in the traditional manner.

The S. Cristina extra-virgin olive oil has a light and fruity aroma. It is lush and sweet to the taste with almond overtones which give an interesting touch of bitterness to the finish. There is very little pepper. This is not an aggressive oil and can be used for almost all culinary applications.

Try it quite simply on Ciabatta bread, mix into salad dressings or serve drizzled over broiled fillets of fish. It is also very good for frying slices of the polenta which is so popular in the Veneto region of Italy. Top with wild mushrooms, also sautéed in the oil with garlic and fennel.

N *Benedetto di Lugana, Peschiera del Garda, Verona, Italy*

950 U.S. gallons

★★ *Extremely good*

Yes

Yes

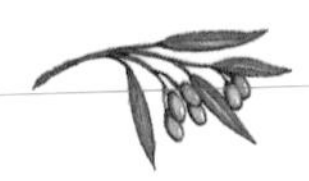

Tenuta
San Guido

This spicy Tuscan extra-virgin olive oil comes from the same estate as the world-famous Sassacaia wine.

The Tenuta San Guido estate is situated just 164 feet above sea level not far from the Tuscan coast near Bolgheri, south of Leghorn. The olive trees account for 270 acres of the 6,100-acre estate, the rest being taken up with vineyards and mixed farming.

The groves include Frantoio, Reggiola, and Moraiolo olives which are picked by hand in November and early December. The olives are then milled on the farm and squeezed at low pressure. The oil remains unfiltered.

The oil has an unusual aroma of olives, apples, and spice with just a hint of liquorice. It tastes grassy-fresh and nutty with a touch of sorrel bitterness and lingering pepper. It is delicious with plain bread, yet it cooks well too. Serve as an addition to chunky soups, pasta sauces, and broiled meats.

Bolgheri, Leghorn, Tuscany, Italy

6,000 U.S. gallons

★★★ *First-class*

Yes

Yes, by appointment

San Guiliano

Originating in a region known as the "Land of Gold," this Sardinian extra-virgin olive oil offers a real depth of delicious fruit.

Three generations of the Manca family have farmed the land here in the northwestern corner of Sardinia near Alghero. The estate includes fields of vegetables as well as the olive groves and much of the produce is bottled and preserved in olive oil.

The groves are planted with Bosese and Frantoiana olives which are hand-picked and traditionally pressed for San Guiliano extra-virgin oil between October and February. The blend is about 70 percent Bosese and 30 percent Frantoiana. This results in an extra-virgin oil with a richly fruity aroma with lemons, apples, and tomato skins. The taste, too, has a real depth of fruit with complex wild flower flavors and light pepper.

The estate also produces a special Fruttato extra-virgin olive oil which is pressed in the traditional manner from olives picked in the first two weeks in November. The blend here is 50 percent of each variety. The oil has a wonderful aroma of tomato and apple skins and a full fruity flavor with green salad leaves, a little pepper, and an elegantly smooth aftertaste.

Alghero, Sassari, Sardinia, Italy

185,000 U.S. gallons in total

★★★ *Excellent*

No

Yes, by appointment

Panzanella Salad

Pour water over two or three thick slices of dry country bread. Squeeze out all the water and chop the bread. Mix with chopped tomatoes, sliced red onions, and plenty of fresh basil. Dress with seasoned extra-virgin olive oil.

Product Box

San Guiliano extra-virgin olive oil
San Guiliano extra-virgin, Fruttato

Olive cream and artichoke cream

Vegetables in olive oil

Frantoio di

Sant'Agata D'Oneglia

Il Capolavoro and Oro Taggiasca

Fragrant extra-virgin olive oils, spicy table olives, exciting jars of Toma cheese, and Muschio mushrooms in olive oil all jostle for attention on the crowded shelves of the company's shop in Imperia.

Generations of the Mela family have lived and worked at Sant'Agata d'Oneglia producing high-quality prize-winning produce of all kinds. Proof of this was their winning of the Delicato section of the 1996 Ercole Olivario competition. They are members of the prestigious Mastri Oleari.

Sant'Agata village is set on a hillside overlooking the Gulf of Imperia and not far from the terraced mountain olive groves. The small and highly-flavored Taggiasca olive is the predominant variety here as in other parts of Liguria.

The harvest lasts from December to February. The fruit is hand-picked or combed from the trees and crushed traditionally with Colombina grindstones dating back to 1827. This is

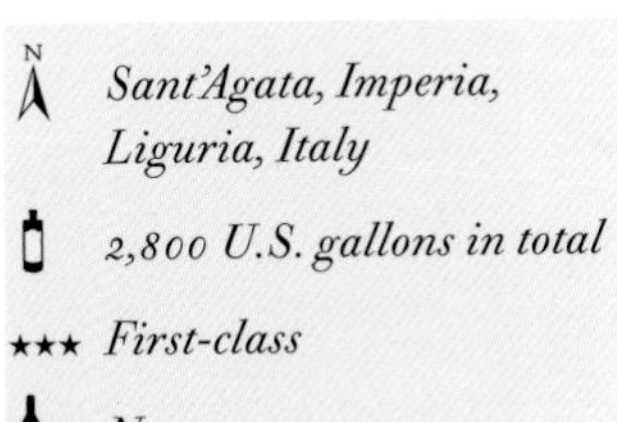

followed by a very light pressing to produce the *Mosto* for the Il Capolavoro oil and normal pressing for the Oro Taggiasca.

The Oro Taggiasco has a very light apple and almond fruitiness with a sweet and delicate aroma. There is very little pepper which makes the oil very attractive to those who do not like a very punchy olive oil.

The company also produce a range of products typical of Ligurian cuisine using recipes which have been in the family for many years.

Product Box

Il Capolavoro extra-virgin olive oil
Oro Taggiasca extra-virgin olive oil

Olives in brine

Green and black olive paste

Pesto Sauce, Ligurian Green Sauce, Arugula Cream and Filbert Sauce

----

Toma Cheese, vegetables, and anchovies in olive oil

Santi

Founded by Carlo Santi in 1843 the Santi wine company has grown enormously, but it has remained true to its traditions of excellence and quality.

The vineyards and olive groves are situated in the beautiful Illasti valley on the eastern shores of Lake Garda. The groves are planted with Casaliva, Favarol, and Leccino olives. The fruit is picked by hand from November to February and pressed in the traditional manner.

Santi extra-virgin olive oil has a light grassy aroma and a good fruit flavor with cut grass and salad leaves. This is followed by some fierce pepper that subsides into a fragrant aftertaste. The overall impression is very sweet.

Gray Mullet

Make deep slashes in each side of the cleaned fish and season all over. Fill with sliced cloves of garlic and fresh bay leaves. Place in a shallow dish and pour on plenty of extra-virgin oil. Leave to marinate for a while. Heat the oven to 325° F. Cover the fish with bay leaves and foil and bake for about half an hour depending on the size of the fish. Remove the foil for the final 10 minutes.

Illasi, Verona, Italy

Not known

★★★ *Excellent*

Yes

Yes, by appointment

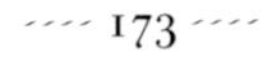

Sartos Fruttato Intenso

This is the first Sardinian extra-virgin olive oil to win the Ercole Olivario Fruttato Intenso category, both in 1994 and 1995.

The olives for this oil come from Seneghe on the west coast of Sardinia near the foot of Monti Ferru, where it drops steeply into the sea. The lush countryside in this part of Sardinia remains virtually untouched.

The Cosseddu family groves at Sartos include Bosana and Terza olives. These are used in a ratio of 70 percent Bosana to 30 percent Terza. The fruit is picked during the first week of November and is pressed in the traditional manner.

Sartos Fruttato Intenso has an intense aroma of grassy fruit with tomato skins and salad leaves. The flavor is very green with cut salad leaves and plenty of pepper. The aftertaste has an interesting flavor very reminiscent of bittersweet sorrel.

Seneghe, Oristano, Sardinia, Italy

5,300 U.S. gallons

★★★ *First-class*

No

Yes, by appointment

Sasso

Exporting to more than 50 countries from three factories in southern Italy, this familiar brand name is now owned by the Nestlé group.

The Sasso company was founded in 1837 by Agostino Novaro who named the business after his wife Paolina Sasso. The company expanded very fast and started to export only a few years after its inception.

Sasso buys in from numerous small and medium-sized growers and producers in the Apulia region of southern Italy. The main olive varieties are Semibitonto, Bitonto, and Andria, but the actual blends are a company secret. The olives are crushed and pressed traditionally.

Sasso extra-virgin olive oil has a sweet lemony aroma with a touch of apples. The flavor is more nutty with light to medium pepper.

Bean Soup

Lightly fry a mixture of onions, garlic, leeks and celery in extra-virgin olive oil. After 4–5 minutes add half a sliced green cabbage, a tin of chopped tomatoes and a can of cannellini beans. Cover with broth and bring to a boil. Season and simmer for 45 minutes. Place a slice of bread in the base of each soup bowl and ladle the soup over.

 *Apulia, Italy*

 *Not known*

★★ *Good*

 *No*

 *Yes, by appointment*

Fattoria

Selvapiana

Once the property of the bishops of Florence, the Fattoria Selvapiana has been in the Giuntini family since 1827. The estate produces excellent wines as well as a first-class extra-virgin olive oil.

Fattoria Selvapiana is situated in the Chianti Rufino region northeast of Florence. The countryside here has changed very little over the centuries and the olives are still grown between rows of grape vines in the ancient manner. About 85 percent of all the olive trees are Frantoio or Coreggiolo. The rest are Moraiolo and Leccino.

Harvesting traditionally starts on 2 November when the olives are still largely unripe. The fruit sits tight on the tree and must be harvested by hand. Once picked the oil is extracted from the olives by the cold centrifugal process.

The oil has a light grassy aroma. The flavor is not at all aggressive but is full of fresh olive flavors with nutty overtones. The pepper builds on the finish but the aftertaste is smooth and elegant. It is a very good oil to serve with steamed vegetables, soups, and salads. Locally it is used as a dip for crudités.

Rufina, Florence, Tuscany, Italy

1,000–1,300 U.S. gallons

★★ *Extremely good*

Yes

Yes

Masi

Serego Alighieri

The estate at Casal Dei Ronchi was established in 1353 by the son of Italy's great poet Dante Alighieri and his descendants live there to this day.

Dante was banished from Florence and lived for a number of years in Verona. His son liked the area so much that he decided to purchase a property and remain in the area. Six hundred years and 20 generations later that property is still in the family.

Casal dei Ronchi is a mixed estate, growing grapes, cherries, and chestnuts as well as olives. It is situated on a small hill near Gargagnago di Valpolicella near Sant' Ambrogio in the Veneto. The main house contains eight apartments which can be rented to use as a center for visiting the region. It is surrounded by vineyards, woods, and gardens.

Above: Olive groves with the Serego Alighieri villa in the background.

Gargagnago di Valpolicella, Verona, Italy

1,300 U.S. gallons

★★ *Extremely good*

Yes

Yes, by appointment

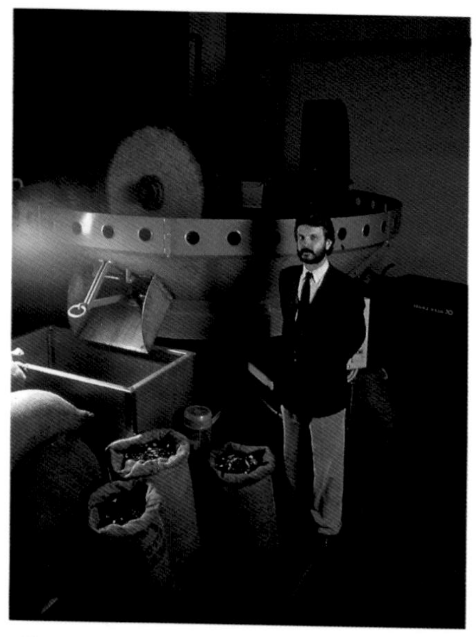

Above: Count Pieralvise Serego Alighieri standing by the "Frantoio".

Four different varieties of olive are grown on the estate. They are Leccino, Grignano, Frantoio, and Casaliva. The fruit is picked by hand toward the middle of November and the olives are milled in the traditional manner. Once pressed the oil is left to settle naturally without being filtered.

This extra-virgin olive oil, which is marketed exclusively by the Masi wine company, has an aroma of warm apples mixed with cut grass. The flavor is sweet and smooth and full of fresh olives with a lightly peppery finish.

Pasta e Fagioli

Finely chop 1 onion, 1 carrot, 1 clove garlic, and 1 stick of celery and cook in extra-virgin olive oil until lightly browned. Add some rosemary, 3 peeled and chopped tomatoes, and 1½ cups freshly shelled broad or lima beans. Cover with 1 cup broth or broth mixed with wine and bring to a boil. Cook over medium heat for 10–15 minutes until the sauce has thickened. Meanwhile cook 2 cups pasta shapes or macaroni until just tender. Toss with the sauce and serve at once.

Elias
Solon

Produced by a company that is now part of a large international conglomerate this brand of Greek extra-virgin olive oil is simple but attractive.

This oil is one of the branded oils produced by Elais, the largest producer of olive oil in Greece. The company, which is based near the port of Piraeus, is now part of the Unilever group. Much of its high-tech production is devoted to mass-market oils and fats but the Solon extra-virgin brand stands out as a reminder of the company's reputation for quality.

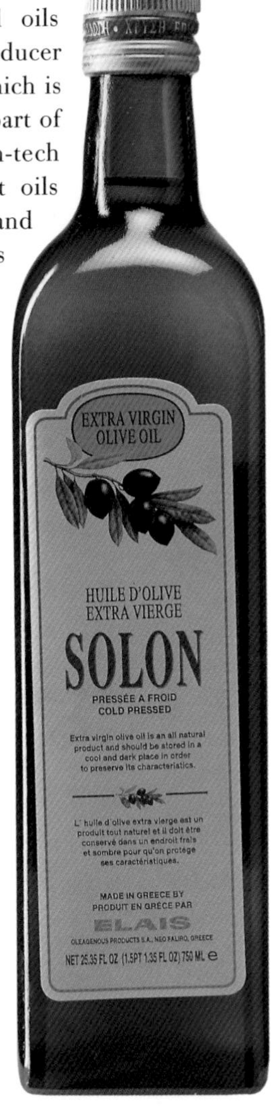

Solon has an aroma of cut hay and mown grass with a touch of lemons and sometimes tomatoes. It is more complex than many Greek olive oils. On the palate it is very sharply grassy with bitter salad leaves and sorrel. Buy this oil at the start of the season and you will be rewarded with a flavor that is sweet and tart at the same time.

Use to make the traditional Greek salad of tomatoes, cucumber, black olives, and Feta cheese or drizzle over kabobs and broiled fish.

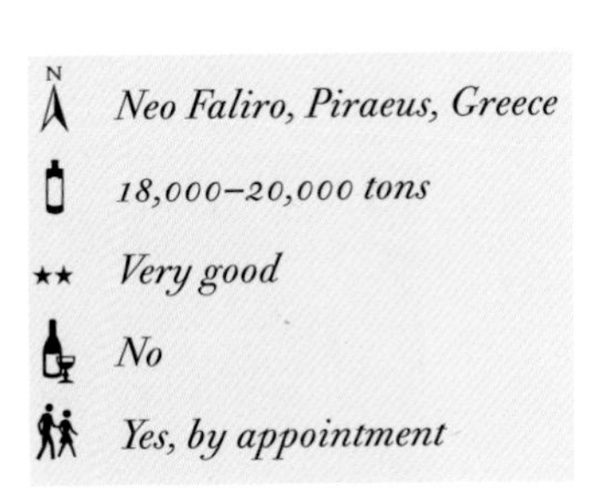

Antico Frantoio da Olive

Sommariva Nuovo Mosto

The Taggiasca olives for this prize-winning extra-virgin olive oil are crushed in a mill which dates back to the sixteenth century.

The Sommariva mill is set picturesquely in the old medieval walls of thirteenth-century Albenga on the Ligurian coast between Imperia and Savona. Domenico Sommariva bought the mill 30 years ago and, with his wife and family, has built up the business in conjunction with their family farm.

As well as operating as a fully functional olive mill from November to March the mill also offers a shop selling a range of local produce, some from the Sommariva's own farm, and a museum of antique equipment both of which are open all the year round.

The Sommariva olive oil is pressed from 90 percent Taggiasca olives and ten percent Pignola and Colombaia. They are grown on the terraced olive groves of the nearby Lerrone and Arroscia valleys. The olives are picked by hand or raked onto nets. The business uses both traditional and centrifugal methods of extraction but

Albenga, Liguria, Italy

Not known

★★★ *First-class*

Yes

Yes

the best oils such as the Nuovo Mosto are crushed and pressed in the traditional equipment at the mill. These oils are unfiltered.

The oil has a pleasant aroma of grass, nuts and apples. The flavor is sweet and attractive with similar tones to the aroma but it can also have a definite flavor of toasted nuts or even chocolate. The oil is quite peppery but with a sweet aftertaste. This is a really interesting oil.

Spaghetti with Potatoes and Pesto

Cook the pasta until al dente. *Drain and toss in extra-virgin olive oil. At the same time cook some firm potatoes. Slice and toss with the pasta and a jar of pesto sauce. Serve with freshly grated Parmesan cheese.*

Product Box

Sommariva Nuovo Mosto extra-virgin olive oil
Sommariva extra-virgin olive oil
Flavored olive oils

Black olives in brine

Olive pastes

A variety of vegetables in olive oil

B. R. Cohn/Olive Hill Company

Sonoma Estate

Only about 200–300 cases of this special extra-virgin olive oil, crushed from French Picholine olive trees imported to California from Provence in the 1870s, are available each year.

The Picholine olive is a rare variety for California but it has found its way onto the B. R. Cohn estate in the heart of the Sonoma Valley. The original owner of the land planted the olive trees on a hill in the center of the estate and the property became known as Olive Hill.

The groves benefit from underground hot springs which results in a microclimate with a temperature about eight degrees above those of surrounding properties. About 75 percent of the olives are fully ripe when they are picked in the last two weeks of November and the first week of December. Once picked the olives are processed by the cold centrifugal method.

The Sonoma estate extra-virgin olive oil is sweet and smooth with a fruity, gently nutty aroma and flavor. There is just a little pepper to add interest to the oil.

The oil is produced and marketed by the Olive Hill Company which was formed in 1992 by Bruce Cohn and his partner Greg Reisinger.

N *Glen Ellen, Sonoma, California, U.S.A.*

Very limited

★★ *Very good*

Yes

Yes

Spectrum Organic

Spectrum is based in California but their organic extra-virgin olive oil comes all the way from Argentina.

Based in Petaluma, Spectrum Naturals produces more than 50 different organic cooking oils, vinegars, and condiments. The organic olive oil is made with 80 percent Arbequina olives. This is a Spanish olive variety which produces a sweetly nutty oil and this one is no exception.

Spectrum organic extra-virgin olive oil has a pleasantly warm and nutty aroma with even more toasted nuttiness in the taste. There is very little pepper, which makes the oil very versatile to use.

This oil makes a particularly good dressing for simple salads of delicate leaves or cooked vegetables. Serve on fresh asparagus with lemon juice or with fried fish fillets.

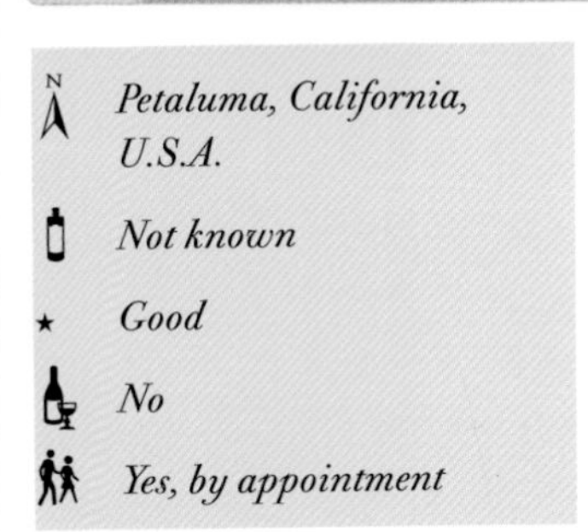

George Skoulikas Ltd

Sunita Organic

This Greek organic extra-virgin olive oil is pressed, unusually, from 100 percent Kalamata olives.

The olives for this oil come from organic olive groves in Mani on the Peloponnese. No artificial fertilizers or pesticides are used on the trees. The olives are milled and pressed in the traditional manner.

The oil, which carries a Soil Association Organic Certificate, has a fresh apple aroma with grassy, lemony tones. The flavor is full and fruity with plenty of mown grass and bitter leaves like watercress. The pepper is deceptively light to start with but builds up to quite a spicy finish.

This U.K. importer also sells a variety of table olives under the same brand name.

Product Box

Sunita extra-virgin olive oil

Sunita green and black Kalamata olives
Sunita French marinated olives with herbs
Sunita Spanish stuffed olives

London, U.K.

Not known

★★ *Extremely good*

No

No

Tenuta di Saragano

The Tenuta di Saragano has been in the family of the Counts of Pongelli since the fifteenth century and the estate has always produced first-class extra-virgin olive oil.

Saragano is situated in the rolling Umbrian hills at Gualdo Gattaneo in the province of Perugia. The groves are planted with Frantoio, Moraiolo, and Leccino olives with Frantoio contributing 60 percent to the blend. The olives are milled and pressed in the traditional manner.

When it is first pressed the Tenuta di Saragano extra-virgin olive oil has an aroma of apples and cut salad leaves. The flavor is all fruit and bittersweet salad leaves. As time passes it often deepens and takes on a more nutty almost chocolaty flavor. There is always good pepper with a sweet fruity aftertaste.

This is an interesting oil to taste over time. Serve with the local ciriole ternana or thick egg-based tagliatelle and sliced garlic or black truffles. It also makes a very good dressing for a salad of arugula with strips of capocollo flavored with wild fennel.

N *Saragano, Gualdo Gattaneo, Umbria, Italy*

2,500 U.S. gallons

★★★ *First-class*

No

Yes, by appointment only

Lorenzo e Luigi di Gandolfo

TORNATORE

This delicately nutty extra-virgin olive oil has a history stretching back to 1700 when the groves at Olivastri in Liguria were bought from the Benedettini friars by Alessandro Gandolfo, founder of the firm.

The friars are credited with bringing the Taggiasca olive variety to Liguria. They organized the olive groves into the now familiar tiers divided by hand built stone walls. The arrangement gives good drainage and makes the land more efficient.

Olivastri nestles in the hills above Imperia. It is sheltered from the mountain winds and because of its height is protected from attack by the olive fly. The soil is not good and the yields are small but the oil wins prizes.

Only Taggiasca olives are grown and the crop goes to produce both olive oil and table olives. There is no choice but to pick by hand in these tiered groves and this is done between December and April. The oil is extracted in the traditional manner and is unfiltered.

Chiusavecchia, Imperia, Liguria, Italy

10,000 U.S. gallons

★★ *Extremely good*

No

Yes, by appointment

The oil has a fresh lemony and attractively nutty aroma. It is usually very sweet and smooth with a touch of grassiness but it can taste even more nutty with toasted undertones and light to medium pepper. It is extremely good drizzled over charcoal-broiled shellfish such as scallops or jumbo shrimp.

Product Box

Tornatore extra-virgin olive oil
Canto organic extra-virgin olive oil
Fascelunghe organic extra-virgin olive oil
Olivastri organic extra-virgin olive oil

Truffle oil
Flavored olive oils

Taggiasca olives in brine

Olive paste and pesto
Anchovy fillets and sun-dried tomatoes in olive oil

UMANI RONCHI

Casal di Serra

The production of this very distinctive extra-virgin olive oil is a minor activity for the well-established Umani Ronchi wine company.

The company, which was purchased by the Bianchi-Bernetti family in 1960, has 250 acres of vineyards in the Verdicchio dei Castelli di Jesi Classico wine producing area of The Marches in central Italy. The olive groves are situated around three villages in Ancona province.

The oil is pressed in the traditional manner from a mixture of Frantoio and Leccino olives which are picked by hand in mid November. The oil has a sweet nutty aroma and flavor with an unusual hint of smoky bacon or prosciutto. It is not very peppery. Few people remain neutral about this distinctive oil – they either love it or hate it!

The earthy quality of this oil makes a wonderful foil for black truffles. Gently heat slivers of the truffle in the oil but do not fry. Pour over your favorite pasta. The oil smooths and rounds out the wonderful flavor of the truffles.

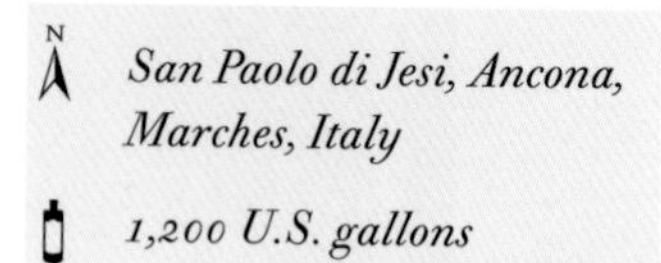

- *San Paolo di Jesi, Ancona, Marches, Italy*
- *1,200 U.S. gallons*
- ★★ *Extremely good*
- *Yes*
- *Yes, by appointment*

Tenuta di VALGIANO

Lucca is often said to produce the best extra-virgin olive oils and this excellent example is as soft and rounded as the local countryside.

Maybe it is the high percentage – around 40 percent – of local olive varieties such as S. Agostino, S. Caterina and Ovo di Accione which give this oil its flowery, herbaceous character. The remaining 60 percent is made up of the usual Tuscan varieties, Leccino, Moraiolo, and Frantoio.

The Tenuta di Valgiano is situated near the village of Valgiano just 12 miles northeast of Lucca. The groves are planted on southfacing hillsides at about 500–800 feet. Picking starts at the beginning of November and goes on until about the middle of December. The olives are crushed in stone crushers and the paste is separated by the cold centrifugal process.

The oil won a silver medal in the "Premio Olio Toscano" awards last year. It has a leafy aroma with artichokes, almonds, and a touch of chocolate. The flavor is rich and attractive with plenty of fruit and medium pepper which creeps up on you. The aftertaste is sweetly nutty.

Valgiano, Lucca, Tuscany, Italy

660 U.S. gallons

★★★ *Excellent*

Yes

Yes, by appointment

YBARRA

This is a good-quality mass-market extra-virgin olive oil marketed by one of Spain's larger producers.

Ybarra was founded back in 1846 by José Maria de Ybarra. Since then it has grown in size and in technical expertise. It is based in Seville in Andalusia and the company claim that its extra-virgin oil is pure Andalusian olive oil. The style of the oil, which is quite sweet and fruity, bears this out.

Ybarra extra-virgin olive oil has an aroma which is strong and warm with a touch of melons in the lemony olive fruit. The oil is fairly heavy but has an attractively sweet finish after medium pepper. It is very suited to everyday use in the kitchen.

Use to make Spanish specialties such as Paella and Gazpacho or in salad dressings and marinades.

SALPICON SALAD DRESSING

Mince a very small onion or shallot, 6 cocktail pickles, 2 tablespoons parsley and a teaspoon of capers. Mix well together. In another container mix ½ cup extra-virgin olive oil with ¼ cup red wine vinegar. Stir in the minced vegetables and a chopped hardboiled egg and spoon over your chosen salad.

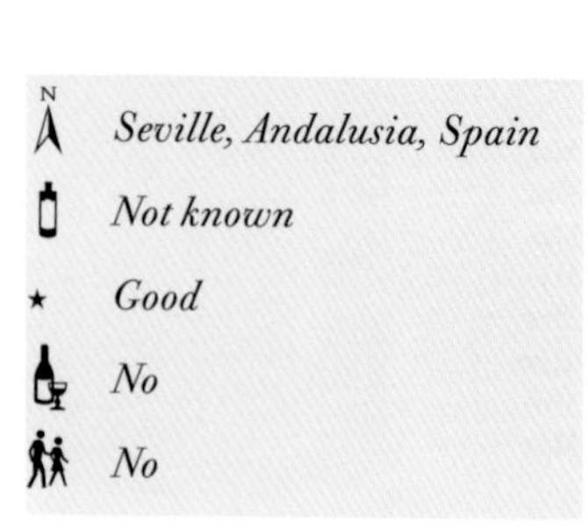

Index

A l' Olivier 45–6
Abbo 42–3
Affiorato Mancianti 44
Agrumolio 149
Alziari 47
American oil production and styles 11, 21, 29
Antinori 48–50
Ardoino 51–2
Argiano 53
Aristeo 54
Athena 55
Avignonesi 56

Badia a Coltibuono & Albereto 57–8
Baena 28, 59
Baggiolino 60–1
Barbi, Fattoria dei 86
Bartolini 62
Berio, Filippo 89
Borelli 63
Borges 64
Bottarelli 65
Brisighella 66

Calaveras 67
Capezzana 68
Carapelli 69
Carbonell 70
Caroli 71–2
Casino di Caprifico 73
Castellare di Ugnana 96
Castello Banfi 75
Castello di Ama 74
Castello Vicchiomaggio 76–7
Colonna 78–80
Columela 80
Cypressa 82–3

Dell'Ugo 84
Donnavascia 85

Eleanthos 82–3

Favarello 85
Felsina 87
Figone's 88
Fontodi 90
Forci 91
Frantoio di Santa Tea 92–3
Frantoio Gaziello 94–5
Frantoio Olive Oil Co. 96
Frantoio Valtenesi 82
French oil production & styles 20, 29
Frescobaldi 97–9
Fresh Olive Company of Provence 100
Fubbiano 101
Fuentebuena 102–3

Gabro 104–5
Gaea 106
Gaziello 94–5
Gemma 107
Gentile di Chieti 148–9
Grappolini 108
Greek oil production & styles 19, 29

Harrison Vineyards 109–10

Iliada 111
Il Lastro 112
Il Poderetto 113–14
Isole e Olena 115
Isnardi 51–2
Italian oil production & styles 18, 28–9

Joseph Foothills 116

Kalamata Gold 132–3
Kydonia 117–18

La Rosa 120
Lake & Co. 119
Laudemio (Antinori) 48–9
Laudemio (Baggiolino) 61
Laudemio (Frescobaldi) 97–9
Laudemio (Pasolini Dall'Onda) 151–2
Le Cesine 121
Le Vieux Moulin 126–7
Lérida 28, 124–6
Les Alpilles 119
L'Estornell 124–6
Lila Jaeger 128
Lungarotti 130–1

Mani 132–3
Mantova 134–5
Mas de la Dame Les Baux 136
Masseria 137
Merchant Gourmet, The 138–9
Meridian Organic 140
Mirum 141
Montevertine 142

Index

Nativo 143
North African oil production 22
Nuñez de Prado 144–5

Oakville Grocery Co. 146
Oleastrum 147
olives & olive oil
- cooking with 32–3, *see also* recipes
- cultivation & harvesting 12–15
- D.O.C. system 26–7
- grades & styles 25–9
- growing areas 16–22
- health & 10, 34–5
- history 7–11
- processing 23–5
- storing 34
- tasting 30–1

Opera Mastra 148–9
Ornellaia 150
Oro de Genave 102–3

Pasolini Dall'Onda 151–2
Phoenician Olive Oil 153
Podere Cogno 154
Poggio Lamentano 155–6
Ponder Estate 157
Portuguese oil production 22
Possenti Castelli 158

Raineri 159
Rainha Santa 160
Ravida 161–2
recipes
- Antinori Panzanella Salad 50
- Bean Soup 175
- Black Truffles 188
- Bruschetta with Anchovies 43
- Bruschetta with Tomatoes 62
- Cheese with Olive Oil 88
- Cream Horns 133
- Fish Baked in Foil 123
- Gazpacho 72
- Gray Mullet 173
- Greek Olive Oil & Hazelnut Cookies 83
- Green Bean & Potato Salad 126
- Green Carpaccio 155
- Lemon Salad Dressing 80
- Marinade for Swordfish or Tuna 55
- Marinated Monk Fish Fillets 140
- Orecchiette with Greens 135
- Olive Oil Cake 118
- Pain d'Aubergine 128
- Panzanelle Salad 170
- Pasta & Fagioli 178
- Pasta with Broccoli 181
- Pinada for Scallops & Shrimp 145
- Pinzimonio 58
- Roasted Fish with Orange-Flavored Olives 149
- Salmon with Barberino Sauce 152
- Salmorejo 145
- Salpicon Salad Dressing 190
- Spaghetti alla Norcina 105
- Spaghettini with Black Olive Paste 139
- Spaghetti with Potatoes & Pesto 181
- Tagliatelle with Arugula & Ricotta 73
- Taramasalata 117
- Warm Chanterelle & Pancetta Salad 110
- White Beans in Olive Oil 61
- Winter Salad 162
- Umbrian Fava Beans 131

Rocca delle Macie 163
Roi 164
Romanico 165

S. Cristina 167
Sadeg Organic 166
San Guido 168
San Giuliano 169–70
Sant'Agate D'Oneglia 171–2
Santa Tea, Frantoio di 92–3
Santi 173
Saragano, Tenuta di 185
Sartos Fruttato Intenso 174
Sasso 175
Selvapiana 176
Serego Alighieri 177–8
single estate oils 16
Solon 179
Sommariva Nuovo Mosto 180–1
Sonoma Estate 182
Spanish oil production & styles 16–17, 28
Spectrum Organic 184
Sunita Organic 184

table olives 14, 17, 19, 20, 21, 36–40
Tomatore 186–7
Triunfo 160

Umani Ronchi 188
Ursini 148–9

Valgiano 189
Vetrice, Fattoria di 112

Ybarra 190